Pascal and Probability

Volume 1 in the "Scientist and Science" series

Enders Anthony Robinson

Professor Emeritus in the
Maurice Ewing and J. Lamar Worzel Chair
Columbia University in the City of New York

2013

Goose Pond Press

Available from Amazon.com and other retail outlets.

Blaise Pascal wrote:

Take away *probability*, and you can no longer please the world;

give *probability*, and you can no longer displease it.

Copyright © 2013

by

Enders Anthony Robinson

Ralph Waldo Emerson told Henry David Thoreau that Goose Pond should be called The Droplet or God's Pond. It was significant to many of Concord's leading literary figures, all of whom walked there often.

Contents

CHAPTER 1. BLAISE PASCAL

Blaise Pascal wrote: "Because," say some, "you have believed from childhood that a box was empty when you saw nothing in it, you have believed in the possibility of a vacuum. This is an illusion of your senses, strengthened by custom, which science must correct." "Because," say others, "you have been taught at school that there is no vacuum, you have perverted your common sense which clearly comprehended it, and you must correct this by returning to your first state." Which has deceived you, your senses or your education?

1.1 Childhood

Blaise Pascal was born at Clermont in Auvergne in France on June 19, 1623. His father, Etienne Pascal, was a well-to-do public official. Blaise's mother, who came of a family of prosperous merchants, died at the age of 30. Blaise was aged three. He had an older sister Gilberte and a younger sister Jacqueline. The children had a governess, but their father Etienne gave them their formal education.

Although sickly, Blaise was a precocious boy. The nature of his illness is not fully known. His father and sisters recognized his exceptional powers. Etienne had his own methods of teaching and he practiced them. Often the instruction was given in the form of conversation. Etienne let natural curiosity guide the direction and emphasis. It was time enough for a child to learn Latin when he was 12, when he could do it more easily. Years later Pascal wrote "By good fortune, for which I cannot be too grateful, I was taught on a peculiar plan and with more than fatherly care." The instruction given by his father was notable for its eccentricity and its rigor. Most of all, the instruction was remarkable for its success.

His father Etienne had an aptitude for mathematics. He decided not to teach Blaise geometry until after he had learned Latin, Greek, history and geography. Mathematics books were not allowed in the house and the subject was not discussed in the daily conversational instructions. However the boy penetrated this barrier on his own. When he was 12, his father came upon him surrounded with diagrams. Blaise was trying to work out the principles of geometry. Etienne forgot his theoretical scruples and was overcome with pride and joy.

When Blaise was eight, the family moved to Paris. Etienne was a sociable man, with a talent for natural science, and he found friends with whom he could share his interests. Etienne gained membership in a group, which included the mathematicians Pere Mersenne, Gilles Personne de Roberval, and Gerard Desargues. Blaise was allowed, while still a child, to accompany his father to the meetings of this group, and was even encouraged to utter his own ideas.

Blaise Pascal's first contribution to mathematics was composed at the age of 16. It was what is now called Pascal's theorem. It states that pairs of opposite sides of a hexagon inscribed in any conic section meet in three collinear points. The theorem was widely praised. Mathematics indisputably became Pascal's great passion.

1.2 First digital calculator

Blaise Pascal wrote: The arithmetical machine produces effects which approach nearer to thought than all the actions of animals. But it does nothing which would enable us to attribute will to it, as to the animals.

In 1640 the Pascal family moved to Rouen. Etienne had been awarded the important but unpopular post of tax assessor. In addition to various administrative chores, Etienne had the task of reassessing the taxes, which meant dealing with seemingly endless columns of figures. His father's drudgery motivated Blaise to invent a machine to perform arithmetic. Pascal began to work on his calculator in 1642, when he was only 19 years old. His calculating machine, named the Pascaline, could add and subtract two numbers directly and multiply and divide by repetition. The device would carry digits from one column to the next.

Different monetary systems, based on the Latin words librae, solidi, and denarii, were widely used in Europe in medieval times. The French system used livres, sols and deniers. There were 20 sols to a livre and 12 deniers to a sol. The English system used pounds, shillings, and pence. There were 20 shillings to a pound, and 12 pence to a shilling. In France, length was measured in toises, pieds, pouces and lignes. There were 6 pieds to a toise, 12 pouces to a pied, and 12 lignes to a pouce. In England, length was measured in fathoms, feet, inches, and lines. There were 6 feet to a fathom, 12 inches to a foot, and 12 lines to an inch. In addition to wheels in base 10, the Pascaline needed wheels in base 6 (for pieds to toises), in base 12 (for deniers to sols, for pouces to pieds, and for lignes to pouces), and in base 20 (for sols to livres).

With the help of local craftsmen, Pascal worked on the idea. Many experimental models were constructed and by 1645 he had a working machine. It was the size of a glove box. By 1652 he had a standard model in production. It was priced 100 livres. Unfortunately his hopes to get rich were not realized.

> Pascal wrote: We owe a great debt to those who point out faults. For they mortify us. They teach us that we have been despised. They do not prevent our being so in the future; for we have many other faults for which we may be despised. They prepare for us the exercise of correction and freedom from fault.

Previous machines were analog. Blaise Pascal had invented the first working mechanical digital calculator, a major innovation whose descendants were to change the world.

Calculators are designed with specific functions such as addition, multiplication, and logarithms. The next major innovation was the Analytical Engine invented by Charles Babbage. The French inventor Jacquard had worked out a method for weaving patterns in rugs as determined by the use of punched cards. The Analytical Engine was intended to use loops of Jacquard's punched cards to control a mechanical calculator, which could use as input the results of preceding computations. The Analytical Engine was never a single physical machine, but rather a succession of designs that Babbage made until his death in 1871. It may be said that the Analytical Engine represented the first mechanical digital computer. The fundamental difference between a digital calculator and digital computer is that a computer can be programmed in a way that allows the program to take different branches according to intermediate results.

Augusta Ada Byron, the countess of Lovelace, was the daughter of the poet Lord Byron. She foresaw the universality of the digital computer.

Using the Analytical Engine as a model, Ada was the one who first devised the notion of a programmable digital computing machine. Babbage focused only on the capabilities of a machine to do routine arithmetical computations. Ada had the genius to foresee the capability of a digital computer to extend beyond numerical calculations. She showed that the potential of a digital computer went into regions far removed from what anyone else had ever realized. Ada emphasized the difference between the Analytical Engine and previous calculating machines. She showed how it was possible to program the Analytical Engine to solve problems of great complexity; problems which otherwise had no association with numbers and mathematics.

> Ada wrote: The Analytical Engine might act upon other things besides number, if objects were found whose mutual fundamental relations could be expressed by those of the abstract science of operations. Supposing, for instance, that the fundamental relations of pitched sounds in the science of harmony and of musical composition were susceptible of such expression and adaptations, the engine might compose elaborate and scientific pieces of music of any degree of complexity or extent. We may say most aptly that the Analytical Engine weaves algebraic patterns just as the Jacquard loom weaves flowers and leaves.

The analysis by Ada was a conceptual leap from previous ideas about the capabilities of computing devices, and foretold the capabilities and implications of the modern computer age. Electronics made the modern computer possible. The history begins with the ENIAC. The ENIAC was the first working electronic digital computer. Work was begun on the Pascaline in 1642. Work was begun on the ENIAC in 1943, which was 301 years later. The Pascaline and the ENIAC are the two breakthroughs that changed the world from analog to digital.

In its working lifetime from 1946 to 1955, the ENIAC was the main computer in the world for the solution of the scientific problems. The ENIAC logged in a total of 80,223 useful hours of operation. The ENIAC was used for many scientific endeavors, including ballistics, atomic energy, weather prediction, wind tunnel design, cosmic ray studies, thermal ignition, and pseudo-random number generation. The ENIAC is reputed to have done more arithmetic calculations than the entire human race had done prior to its construction.

1.3 Pascal's principle

Pascal wrote: You see, if the height of the mercury column is less on the top of a mountain than at the foot of it (as I have many reasons for believing, a though everyone who has so far written about it is of the contrary opinion), it follows that the weight of the air must be the sole cause of the phenomenon, and not that abhorrence of a vacuum, since it is obvious that at the foot of the mountain there is more air to have weight than at the summit, and we cannot possibly say that the air at the foot of the mountain has a greater aversion to empty space than at the top.

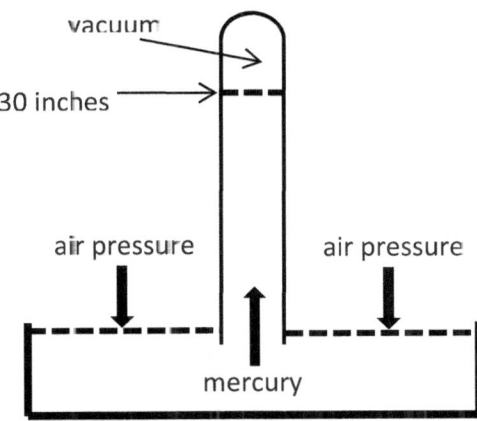

Figure 1. Torricelli's barometer

Let us explain the above passage from Pascal. Evangelista Torricelli was a friend and disciple of Galileo. In 1643 Torricelli took a tube approximately 40 inches long that was sealed at the top. He filled it with mercury. He put his thumb over the open end and turned the tube upside down. Now the sealed end was at the top and the open end at the bottom. He then placed the tube in an open bowl of mercury and took his thumb away. The mercury in the tube sank downward until its level was 30 inches above the level in the bowl. As a result, a vacuum was created in the tube down from the sealed end to the level of mercury in the tube. In this way, Torricelli became the first one to create a sustained vacuum. The height of the mercury column fluctuated with changing atmospheric pressure. Torricelli had invented the barometer.

The young Blaise Pascal was captivated by the work of Torricelli. Pascal realized that it was the weight of the atmosphere pushing done on the mercury on the bowl that sustained the 30 inches of Mercury in the tube. Fish live at the bottom of a sea of water, and that water has weight. Pascal realized that people live at the bottom of a sea of air and that air has weight. The mercury in the tube did not fall into the bowl but sank only part of the way down the tube. Pascal asserted that the mercury stands in a column high enough to make equilibrium with the weight of the external air which forces it up. With glass tubes of different lengths and breadths, Pascal established that the space vacated by the falling liquid was indeed a void. For a given liquid the height of the column in the tube is always the same. It is not affected by the space at the top. It is not sucked up from above but is pressed up from below. Pascal's results were conclusive. He had established the disciplines of hydrostatics and hydrodynamics. He wrote two treatises: "The Great Experiment on the Equilibrium of Liquids" and "On the Weight and Mass of Air," both of which were published posthumously.

Pascal's experiments led him to principle of transmission of pressure in fluids. A fluid is a substance, as a liquid or gas, that is capable of flowing and that changes its shape at a steady rate when acted upon by a force tending to change its shape. Pressure exerted at any place on a fluid in a closed vessel is transmitted undiminished throughout the fluid and acts at right angles to all surfaces. Suppose two tanks of equal size are filled with water and joined by a pipe at the base. The water in the tanks will be in equilibrium, like equal weights in the pans of a scale. But suppose one tank is 100 times larger than the other? The water will still be in equilibrium.

Pascal, in his own clear wording, writes, "If a vessel full of water, closed on all sides, has two openings, the one a hundred times as large as the other, and if each be supplied with a piston which fits exactly, a man pushing the small piston will exert a force which will equilibrate that of a hundred men pushing the piston which is a hundred times as large, and will overcome that of ninety-nine. And whatever may be the proportion of these openings, if the forces applied to the pistons are to each other as the openings, they will be in equilibrium. Whence it appears that a vessel full of water is a new principle of mechanics, and a new machine for the multiplication of force to any required degree, since one man will by this means be able to raise any given weight."

In other words, Pascal's principle says that given a fluid in a totally enclosed system, a change in pressure at one point in the fluid is transmitted to all points in the fluid, as well as to the enclosing walls. If you have a fluid enclosed in a pipe and if you change the pressure in the fluid at one end of the pipe, then the pressure changes all throughout the pipe.

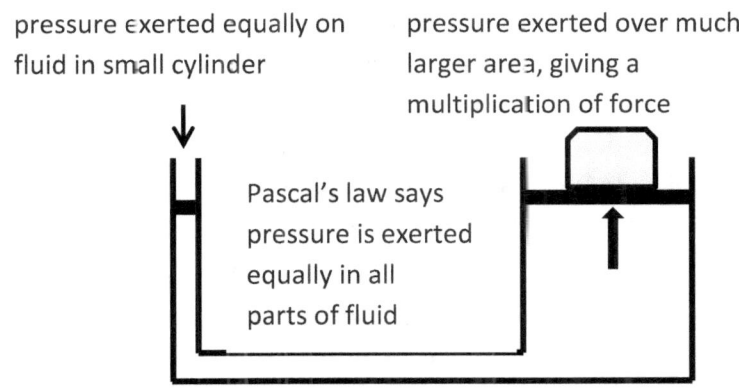

Figure 2. A hydraulic lift magnifies force

The figure shows a system of enclosed fluid with two hydraulic pistons. The piston on the left has a small head of area A_1 and the piston on the right has a large head of area A_2. The force in the small cylinder must be exerted over a much larger distance. In other words, a small force over a large distance is traded for a large force over a small distance. If force F_1 is applied to the left piston, what is the force F_2 on the right piston? Pressure at each point is the ratio of force over area. According to Pascal's principle, the pressure is the same everywhere inside the fluid, so

$$\frac{F_1}{A_1} = \frac{F_2}{A_2}$$

This means that a large force can be developed from a small force if the ratio of the piston sizes is big. For example, say the area of the right piston is larger than the left piston by a factor of 100. In such a case, any force applied to the left piston will be multiplied by 100 times on the right piston. By using a small piston on one side and a large piston on the other side, huge forces can be created. However, you do not get something for nothing. The smaller piston must be pushed 100 times as far as the larger piston will move. The hydraulic lift is similar to a lever, where a small force applied through a large distance can move a heavy object a small distance.

1.4 Probability

In 1653 the Chevalier de Mere, a gambler, asked Pascal to solve two problems of practical use. Pascal scientific interests were stimulated. The first problem was: "When one plays with two dice, what is the minimum number of throws on which one can advantageously bet that a double six will turn up?" Pascal gave the solution: 24 throws would be a bad bet; 25, a good one. (See Problem 22, Chapter 14.)

The other question was: Two players have agreed that the stakes will go to the one who first wins three games. Before this happens, the play is interrupted. How, under different circumstances, are the stakes to be divided? Pascal found an ingenious and simple answer. (See sections 3.1 and 3.2 in Chapter 3). He sent his solution to Pierre de Fermat in Toulouse, who had got the same results by algebraic methods. Pascal was very pleased. "I see," he wrote, "that the truth is the same in Toulouse and in Paris." The independent work of both men laid the basis of the mathematical theory of probability.

Pascal's success in solving Chevalier de Mere's problems stimulated him to further study of the mathematical theory of chance. One of his brilliant ideas was the arithmetical triangle, now known as Pascal's triangle. (See section 8.1 in Chapter 8 and section 16.3 in Chapter 13.) Pascal wanted to reduce the incertitude of chance to an exact art, with the rigor of mathematical demonstration, thus creating a new science which could claim the title: "The geometry of hazard."

In his later years, Pascal turned his attention towards religion. Although religious questions were uppermost in his thoughts, he neither entirely abandoned science nor renounced fashionable society. In 1659, Pascal "bade a final farewell to the glories and quarrels of science." Once more he wanted to meet with his friend Fermat, but he was too weak to undertake the journey. At the time Pascal and Fermat were the most distinguished mathematicians of Europe. Today they rank among the greatest mathematicians of all time.

Pascal wrote to Fermat saying, "I find geometry the noblest exercise of the mind, yet I know it to be so useless that I see no difference between a geometer and a clever artisan. I call it the loveliest occupation in the world, but only an occupation. A singular chance about a year or two ago did set me at mathematics, but having settled the matter I am not likely ever to touch the subject again."

Pascal was dying. He had resigned himself and was waiting upon God, writing pages of his "Pensees. On August 18, 1663 he died, aged 39. Pascal's scientific achievements were extraordinary, but the "Pensees" was his supreme achievement. No one has yet transcended his insight into man's condition.

1.5 Pensées

Blaise Pascal wrote: There are then two kinds of intellect: the one able to penetrate acutely and deeply into the conclusions of given premises, and this is the precise intellect; the other able to comprehend a great number of premises without confusing them, and this is the mathematical intellect. The one has force and exactness, the other comprehension. Now the one quality can exist without the other; the intellect can be strong and narrow, and can also be comprehensive and weak.

At his death, Pascal left numerous pages of handwriting, grouped in a tentative order. In 1670, these notes were arranged and printed as a book titled "Pensées de M. Pascal sur la réligion, et sur quelques autres sujets." This book "Pensées" soon became a classic. It was Pascal's most influential work. It is universally considered to be a masterpiece. It represents a landmark. It has been praised as "the finest pages in the French language "and as "the most eloquent book in French prose." In the book, Pascal discusses philosophical paradoxes. Included are topics such as infinity and nothing, faith and reason, soul and matter, death and life, meaning and vanity. Pascal gives the most penetrating analyses, arriving at no definitive conclusions besides humility, ignorance, and grace. Pascal's "Pensées" takes its place among the most profound and beautifully written works of genius in history.

"Pascal's Wager" is found in the "Pensées." In devising the system incorporated in Pascal's Wager, Pascal originated the discipline known today as decision theory; that is, the making of a decision under uncertainty. The wager applies decision theory to the belief in God. Pascal argues that it is always a better "wager" to believe in God than not to believe in God. Pascal shows that the expected value gained from belief in God is always greater than the expected value resulting from non-belief. Pascal's argument does not deal with the existence of God. Instead it is an argument for the belief in God. With his wager, he sought to demonstrate that believing in God is advantageous to not believing.

> Pascal wrote: Belief is a wise wager. Granted that faith cannot be proved, what harm will come to you if you gamble on its truth and it proves false? If you gain, you gain all; if you lose, you lose nothing. Wager, then, without hesitation, that He exists.

Pascal's wager has the following possibilities:

	God exists	God doesn't exist
You may believe in God	You go to heaven; your gain is infinite	Your loss is finite and therefore negligible.
You may not believe in God	You will go to hell; your loss is infinite.	Your gain is finite and therefore negligible

In this way, Pascal deduced that it would be better to believe in God unconditionally. This argument is valid within its assumptions. Pascal had invented "game theory."

In general terms, mathematics can be divided into four main divisions representing quantity, structure, space and change:

Arithmetic	Algebra	Geometry	Infinitesimal analysis
quantity	structure	space	change

In Pascal's time, infinitesimal analysis had not been invented. Arithmetic and Algebra had been in practical use for centuries, but were still far from ultimate development. Only that part of Geometry, as put forth by Euclid, could be considered to be in a developed state.

Pascal distinguishes

1. The logical (or mathematical) type of mind which deduces everything by mathematical reasoning

2. The intuitive type of mind which sees everything at a glance

Blaise Pascal wrote: Man is but a reed, the most feeble thing in nature; but he is a thinking reed. The ent re universe need not arm itself to crush him. A vapor, a drop of water suffices to kill him. But, if the universe were to crush him, man would still be more noble than that which killed him, because he knows that he dies and the advantage which the universe has over him; the universe knows nothing of this.

All our dignity consists, then, in thought. By it we must elevate ourselves, and not by space and time which we cannot fill. Let us endeavor, then, to think well; this is the principle of morality.

CHAPTER 2. BACKGROUND IN LANGUAGE

Blaise Pascal wrote: I have spent much time in the study of the abstract sciences; but the paucity of persons with whom you can communicate on such subjects appalled me. When I began to study man, I saw that these abstract sciences are not suited to him, and that in diving into them, I wandered farther from my real object than those who knew them not, and I forgave them for not having attended to these things. I expected then, however, that I should find some companions in the study of man, since it was so specifically a duty. I was in error. There are fewer students of man than of geometry.

2.1 Mathematical mind and intuitive mind

Prehistoric people lived in a word of uncertainty, and their language reflected their condition. The Proto-Indo-European language had four moods: (1) indicative, (2) imperative, (3) subjunctive, and (4) optative. The subjunctive mood is used to express states of unreality such as wish, possibility, judgment, or opinion. The optative mood is a grammatical mood that indicates a wish or hope. However, over time, most Indo-European languages lost the optative entirely, or incorporated optative forms into the subjunctive. Some Germanic verb forms often known as subjunctives are actually descendants of the Proto-Indo-European optative. Likewise in Latin, the newer subjunctive is based on the Indo-European optative.

The subjunctive and optative do not refer directly to what is necessarily real. They commonly refer to things that have not yet occurred. For this reason, the subjective mood and optative mood are probabilistic in nature. Each time prehistoric hunters went into field or forest, they subjectively had to evaluate the probability of finding game. They well knew how to use the subjunctive and optative in their speech.

With "intuitive intelligence," mankind used the nuances of language to express probabilistic ideas. In this role, the subjective and the optative were essential. Ancient and medieval laws of evidence developed a grading of degrees of proof, probabilities, and presumptions to deal with the uncertainties of evidence in court.

In 1654, Antoine Gombaud, the Chevalier de Mere, a wealthy French nobleman, wanted to find the reason why he consistently lost money in a certain game of dice. Until then, gamblers would blame losses on bad luck. Not so with the Chevalier de Mere. He posed the question to Pascal, who solved the problem with mathematics. Pascal started a correspondence with Pierre de Fermat on the subject. Because of their correspondence, they are recognized as the originators of the theory of probability.

In Paris, Christiaan Huygens was told of their correspondence, which was in an unorganized form. In 1657, Huygens wrote a treatise that gave the first comprehensive treatment of probability theory. In 1662 Sir Robert Moray sent to Christiaan Huygens the life table compiled by John Graunt. From the life table, Christiaan Huygens and his brother Lodewijk Huygens originated the concept of "life expectancy," which was the beginning of the discipline of mathematical statistics.

Pascal was the first to use mathematics to solve probabilistic problems. With "mathematical intelligence," mankind could now use the nuances of symbols and equations to express probabilistic ideas. It was a breakthrough that was essential to the development of the modern world. Pascal was also the first to invent a mechanical digital calculating machine that worked. It was manufactured and sold on the open market. This invention, a result of mathematical intelligence, led to the modern digital computer.

For thousands of years people have been faced with uncertainty. It is part of the human condition. It is reflected in the words and the structure of language. Before Pascal, knowledgeable people by necessity had to rely on the intuitive mind in their daily lives. They used the words of language. After Pascal, knowledgeable people could also make use of the mathematical mind, if they were so motivated. In this chapter, we will deal with the words that impart probabilistic meaning. The mathematical theory of probability will be developed in the following chapters.

Pascal introduced the distinction between the mathematical mind (esprit géométrique) and the intuitive mind (esprit de finesse). This distinction is important. In the mathematical mind, the principles are palpable, but expressed in symbols and equations, so that it is difficult to turn one's mind in that direction. However, in the intuitive mind, the principles are found in the common use of words. The principles stand in front of the eyes of everybody. The vision of a person must be good because the principles are subtle and numerous, so that it is most likely that some will escape notice.

Few if any writings have been intellectually more influential than the passages in the Pensées, where Blaise Pascal writes on "The difference between the mathematical and the intuitive mind."

> In the mathematical mind, the principles are palpable, but removed from ordinary use; so that for want of habit it is difficult to turn one's mind in that direction: but if one turns it thither ever so little, one sees the principles fully, and one must have a quite inaccurate mind who reasons wrongly from principles so plain that it is almost impossible they should escape notice.

> But in the intuitive mind the principles are found in common use, and are before the eyes of everybody. One has only to look, and no effort is necessary; it is only a question of good eyesight, but it must be good, for the principles are so subtle and so numerous, that it is almost impossible but that some escape notice. Now the omission of one principle leads to error; thus one must have very clear sight to see all the principles, and in the next place an accurate mind not to draw false deductions from known principles.

> All mathematicians would then be intuitive if they had clear sight, for they do not reason incorrectly from principles known to them; and intuitive minds would be mathematical if they could turn their eyes to the principles of mathematics to which they are unused.

> The reason, therefore, that some intuitive minds are not mathematical is that they cannot at all turn their attention to the principles of mathematics. But the reason that mathematicians are not intuitive is that they do not see what is before them, and that, accustomed to the exact and plain principles of mathematics, and not reasoning till they have well inspected and arranged their principles, they are lost in matters of intuition where the principles do not allow of such arrangement. They are scarcely seen; they are

felt rather than seen; there is the greatest difficulty in making them felt by those who do not of themselves perceive them.

These principles are so fine and so numerous that a very delicate and very clear sense is needed to perceive them, and to judge rightly and justly when they are perceived, without for the most part being able to demonstrate them in order as in mathematics; because the principles are not known to us in the same way, and because it would be an endless matter to undertake it. We must see the matter at once, at one glance, and not by a process of reasoning, at least to a certain degree.

And thus it is rare that mathematicians are intuitive, and that men of intuitior are mathematicians, because mathematicians wish to treat matters of intuition mathematically, and make themselves ridiculous, wishing to begin with definitions and then with axioms, which is not the way to proceed in this kind of reasoning. Not that the mind does not do so, but it does it tacitly, naturally, and without technical rules; for the expression of it is beyond all men, and only a few can feel it.

Intuitive minds, on the contrary, being thus accustomed to judge at a single glance, are so astonished when they are presented with propositions of which they understand nothing, and the way to which is through definitions and axioms so sterile, and which they are not accustomed to see thus in detail, that they are repelled and disheartened.

Mathematicians who are only mathematicians have exact minds, provided all things are explained to them by means of definitions and axioms; otherwise they are inaccurate and insufferable, for they are only right when the principles are quite clear.

And men of intuit on who are only intuitive cannot have the patience to reach to first principles of things speculative and conceptual, which they have never seen in the world, and which are altogether out of the common.

2.2 Hap, luck, chance, fortune, lot, hazard

In this book we do not wish to undertake to produce a compendium of common word usage. Instead we wish to develop certain concepts on a mathematical basis with a view toward their intuitive background and their applications. Often, the content of the concepts are derived from the meanings popularly given to the words. The correctness of these concepts can only be judged from their fruitfulness, both as creative ideas and as useful instruments for advancing scientific investigation.

Uncertainty is a part of life. Let us look at its manifestation in language. The following abbreviations are used: adj. = adjective, adv. = adverb, n. = noun, v.i. = verb intransitive, v.t. = verb transitive.

An English word whose origin is the Old Norse word "happ" is "hap." Hap (n.) means luck or chance, whereas to hap (v.i.) means to befall. From this basic word the following English words are derived:

to happen (v.i.): to occur by chance.

happening (n.): occurrence, that which happens.

haply (adv.): by chance.

haphazard (adj.): random, determined by chance.

perhaps (adv.): by some chance, maybe.

happy (adj.): favored by luck or fortune.

happily (adv.): by good fortune.

happiness (n.): good luck, a state of well-being.

happy-go-lucky (adj.): trusting to luck, easygoing.

hapless (adj.): unlucky.

mishap (n.): ill luck, an injurious accident.

The six words (n.)

| hap | luck | chance |
| fortune | lot | hazard |

cannot be properly defined. They are more or less synonymous. They all agree in designating that which happens, either partially or entirely as the result of unknown, unconsidered, or unpredictable forces. Synonymous words such as these continually grope their way through language by connotation; that is, by their suggestive significance. There is nothing less precise than connotation which changes from person to person and from time to time. It is based on associations, many of which are below the conscious level. However, without connotation we could not have language as we know it.

The word "luck" (n.) usually refers to that which happens to one personally or individually. The word luck is associated etymologically and by continual use with gambling, as

> It was a wonderful run of luck.
>
> Don't abuse luck by playing all over the roulette table layout.
>
> Playing the roulette wheel is strictly luck and nothing else.
>
> It would be a kick in the teeth to the luck which had been given him.
>
> Luck only enters the game at sporadic intervals.
>
> Luck has no influence in the long run.

The word luck is used constantly in colloquial speech, as

> His winning was just dumb luck.
>
> It was just his luck to fall off the boat.
>
> He had bad luck at his job today.
>
> The hunter had good luck.
>
> The man tried his luck at the old mine.

The word luck unqualified can imply a happy outcome, as

> Wish me luck.
>
> He had luck.

The derived word "lucky" (adj.) means favored by luck, fortunate, as

> My lucky number is 3.
>
> I have no lucky numbers.
>
> He was lucky to win again.
>
> You were lucky to find your lost ring.
>
> Lucky in cards; unlucky in love.

In the Germanic languages we have this correspondence

English	luck	lucky
Swedish	lycka	lycklig
Danish	lykke	lykkelig
Norwegian	lykke	lykkelig
Icelandic	lukka	Lukkulegur
Dutch	geluk, luk	gelukkig

German   Gluck   gelucklich

An American would say good luck. A Frenchman for the same occasion would say bonne chance. The English word "chance" (n.) comes from the Old French "cheance." In turn, "cheance." comes from the Late Latin "cadentia" which means a falling, especially of dice or of fortune. The Late Latin word "cadentia" itself is derived from the Latin word "cadere" meaning to fall or to happen. Thus we have found an important link: the word chance has its roots with the falling of dice.

The English word "chance" (n.) serves often as the general name for the incalculable and fortuitous element in human existence and in nature as

Most phenomena are influenced by chance.

In common usage, the word chance seldom loses implications derived from its original association with the throwing of dice and the selection of one outcome out of many possible outcomes by this means. Consequently, it may mean determination by irrational, uncontrollable forces, as

He left things to chance.

It may mean any one of the contingencies on which a player takes a risk in a game of chance, as

Number 7 is my chance.

It may mean any risk or gamble, as

This farm is my chance.

Take a chance and accept the position.

It is my only chance.

It may mean an opportunity that comes seemingly by luck or accident, as

When I get a chance, I'll leave.

It was my last chance to escape.

He has long hoped for a chance of promotion.

It may mean a possibility of something happening, as

There is little chance of rain.

The chances are even.

It may mean a possibility of success among many possibilities of failure, as

> He is always willing to take a chance.
>
> What are his chances?
>
> The chances are against him.
>
> He has one chance out of a million.

It may mean a measure that an event happens, as

> What is the chance of the bullet hitting the target?
>
> Its chance of happening is 2 out of 10, or 0.2
>
> What is the chance of drawing two kings in a poker hand?
>
> What chance is there of rain this weekend?
>
> There is little chance.

The English word "chance" (v.i.) means to happen, come, or arrive without design or expectation, as

> He chanced upon this new concept.
>
> Chance (v t.) means to take the chances of, to risk, as
>
> To chance a swim in such cold water is foolish.

A word etymologically related to chance is "case." Case (n.) comes from the Old French "cas," which is from the Latin "casus," which in turn is from the Latin "cadere" meaning to fall, to happen.

The English word "chance" (adj.) means happening by chance. Synonymous are "haply," "accidental," "fortuitous," and "aleatory."

A word related to chance is "accident." The word accident came into English through French from the Latin word "accidere" which means to happen and which is a compound of "ad" plus "cadere" (to fall). Thus an accident is a happening or an event which takes place without one's foresight or expectation, and now often means one of an afflictive, injurious, or unfavorable character, as

> He was in an automobile accident.

The derived word "accidental" (adj.) means happening by chance or unexpectedly. Although the usage of the word accidental usually stresses chance, it now sometimes may stress non-essentiality.

The word "fortune" (n.) came from Old French "fortune," which in turn came from Latin "fortuna." In addition to its meaning of luck, it also means (sometimes capitalized) the personified power of chance, the personification of the cause of that which befalls in a sudden or unexpected manner. Thus we hear about Dame Fortune who is the same mysterious woman as Lady Luck. Sometimes she is pictured as having a large wheel which she spins to determine the various happenings of life.

The derived word "fortunate" (adj.) means lucky. The three synonyms "lucky, fortunate, happy" all connote having a favorable issue. "Lucky" implies success by chance rather than as the result of merit. "Fortunate" is less suggestive of a favorable accident and may carry the connotation of being watched over by a higher power or of being favored beyond one's desserts. "Happy" combines the implications of lucky and fortunate to express gratification in the sense of well-being and of complete satisfaction. Thus happy becomes synonymous with "glad" (from Anglo-Saxon "glaed" meaning "bright, glad") to mean characterized by joy or pleasure. The English word glad is the same as the Swedish word glad, both originating from the same Teutonic word. Whereas in English we have seen that happy has become synonymous with glad, in Swedish it has turned out that lycklig has become synonymous with glad.

The English word "fortuitous" (adj.) comes from the Latin word "fortuitous," which is from the Latin "forte" (adv.) meaning by chance, which in turn is from the Latin "fors, fortis" meaning chance. Fortuitous means happening by chance or accident and is synonymous with accidental. Whereas both accidental and fortuitous mean "not expected, outside of the regular course of things," fortuitous more strongly suggests chance than accidental and often connotes absence of a cause.

The word "lot" comes from the Anglo-Saxon word "hlot," which was an object used as a counter or check in determining a question by chance. To choose by lot is the use of such a counter as a means of deciding something. Lot has come to mean that which befalls one from a choice by lot, and hence a share or allotment. Lot has also come to mean hazard, fortune, especially the fate which falls to one by the will of an overruling power, as

He is a man content with his lot.

Always in the usage of lot to mean the state or end predetermined for one, there is the suggestion of the operation of blind chance. Like the word luck, the word lot usually refers to that which happens to a person as an individual.

The derived word "lottery" (n.) means a scheme for the distribution of prizes by lot, especially such a scheme in which lots, or chances, are sold, and has come to mean figuratively an affair of chance .

The word "hazard" (n.) comes to English through the Old French word "hazard" which is from Arabic al-zahr which means the die. Originally hazard was a dice game of which the present day craps is a simplified form. Thus hazard came to mean chance. In addition, hazard also has come to mean "risk, danger, peril."

The derived word "hazardous" (adj.) thus means both "depending upon chance or luck" and "dangerous, risky." In fact hazardous implies so many chances of evil and/or harm that the thing so described is exceedingly dangerous.

The Latin word "alea" meaning "die, chance" did not come directly into the English language. Nevertheless, this word is known from the famous expression of Caesar when he crossed the Rubicon River:

Alea jacta est (which means the die is cast).

The derived Latin word "aleatorius" has come into English as "aleatory" (adj.). It means "pertaining to or resulting from luck." In legal terminology, it means depending upon an uncertain event or contingency as to both profit and loss. Thus aleatory contracts include lottery agreements, wagering contracts, and insurance contracts.

2.3 Random.

Let us now look at the word "random." Its earliest known meaning is that of furious action, such as a charge of cavalry. It seems to be connected with the Teutonic word "rand" which means brim, and implies the furious and irregular action of a river full to the brim. The

Swedish word "rand" means "verge, brink, brim." The English word random is related to the Old French "randon," which means violence, rapidity. Random (n.) means an aimless course or progress. It is used chiefly in the expression "at random," which means "without definite aim, direction, rule, or method." At random and at haphazard are synonymous, in the sense now described.

Random (adj.) and haphazard (adj.) are synonyms. Both mean having a cause or a character that is determined by accident rather than by design or by method. Random signifies that there is no fixed or clearly defined aim, purpose, or evidence of method or system or direction, and hence implies no or little guidance by a governing mind, eye, objective, or the like, as

The man aimed the bow at random.
The knight wandered at random.

Here random indicates the aimless character of the performance, as contrasted to the definite intention to hit a certain mark. Some examples of this usage are

He only has a random collection of books on science.
It was only a random allusion to the subject.
It was a random shot into the woods.

Haphazard on the other hand signifies an aimlessness, or randomness, that is more or less at the mercy of chance, as

The garden was a haphazard arrangement of shrubs and plants.
The rain fell in a haphazard pattern on the lawn.
The foreign policy is haphazard.

In mathematics the word random is used as the word haphazard is used in ordinary language, that is, with a definite connotation of chance.

2.4 Probable and likely

In this section we want to look at the English words "probable" (adj.) and "likely" (adj.) and at the corresponding words "probability" (n.) and "likelihood" (n.).

First let us look at the correspondence

Latin	probabilis
English	probable
French	probable
Italian	probabile

The Latin word "probabilis" (adj.) comes from the Latin word "probare" meaning to try, to test, to prove. The English word probable comes through French from the Latin "probabilis." Thus etymologically probable is related to provable.

The Latin word "verisimilis" (adj.) comes from the Latin "verus" true (genitive: "veri") + "similis" like. The following correspondence exists.

Latin	verisimilis	veri true + similis like
English	verisimilar, likely	veri true + similar like
German	wahrscheinlich	wahr true + scheinlioh appearing
Dutch	waarschijnlijk	waar true + schijnlijk appearing
Swedish	sarnolik	sanno true + lik like
	trolig	tro true + lig like
Danish	sandsynlig	sand true + synlig appearing
	trolig	tro true + lig like
Norwegian	sandsynlig	sand true + synlig appearing
	trolig	tro true + lig like)
Icelandic	liklegur	lik like + legur ly)
Russian	veroyatnii	vernii true
French	vraisemblable	vrai true + semblable similar
Italian	verisimile	veri true + simile like

All the above words were built up from the basic stems to express the meaning: true-like, likely, true-seeming, true-appearing, true-resembling, true-similar.

The English word "likely" (adj.) is made up of "like" (adj.) (which means having the same, or nearly the same, appearance, qualities, or characteristics) + "ly" (which means like in appearance, manner, or nature). Like comes from Anglo-Saxon "lic" body, and it originally means having the same body or shape), whereas -ly also comes from Anglo-Saxon "lic" body). In Swedish both usages of lik are preserved, in that the Swedish word lik (n.) means body, corpse, and the Swedish word lik (adj.) means like.

The Scandinavian words "sann , sand" have the same root as the English word "sooth" (Anglo-Saxon "soth," true). (Note that "sanno" in Swedish is the genitive form of "sann" .)

The Scandinavian suffix "lig" as well as the Icelandic suffix "legur" correspond to the English suffices "ly, like".

The English words "probable" and "likely" are synonyms. They signify uncertainty that may be, or become, true, real or actual. Something is probable if it has so much evidence in its support or seems so reasonable that it commends itself to the mind as worthy of belief, although not to be accepted as a certainty. Thus the probable conclusion from evidence at hand is the one which the weight of evidence supports even though it does not provide proof, as

He is the probable author of the medieval manuscript.

The probable cause of the fire was faulty wiring.

That are the probable expenses for a trip to Europe ?

What is the probable origin of the folk song ?

The word "likely" is very close to the word "probable", and so they can often be used interchangeably. In contrast to probable, likely does not as often or as invariably suggest grounds sufficient to warrant a presumption of truth, as

It is not the probable place where they will meet, but it is still likely.

Also "likely" is sometimes used in the sense of promising because of appearances, ability to win favor, etc., as

She is a likely candidate.

Using these words we may establish a scale of expressions, as

It is certain to snow.

It is very likely to snow.

It is likely to snow.

It is as likely to snow as not .

It is unlikely to snow.

It is very unlikely to snow.

It is certain not to snow.

or as

He is the certain winner.

He is the probable winner.

He is a likely winner.

He is an unlikely winner.

The word "likelihood" (n.) is derived from likely. We have the correspondence

English	verisimilitude	
	likelihood	
German:	Wahrscheinlichkeit	(keit hood)
Dutch:	waarschijnlijkheid	(heid hood)
Swedish:	sannolikhet	(het hood)
Danish:	sandsynlighed	(hed hood)
Norwegian	sandsynlighed	(hed hood)
Icelandic	l kindi	
Russian:	veroyatnost	
French:	vraisemblance	
Italian:	verisimiltudine	

The English word "probability" (n.) is derived from probable. The corresponding word in French is "probabilite" and in Italian "probability."

"Likelihood" and "probability" are synonyms. They mean quality or state of being likely or probable, as

The likelihood of winning is small.

His probability of being chosen is great.

The probability of hitting the target is small.

As we have seen, the word "chance" (n.) etymologically is related to the falling of a die. On the other hand, the word "probability" (n.) etymologically is related to trying or testing a thing against some truth. In mathematical usage, the words probability and chance are overlapped so that they both signify a measure that an event happens.

2.5 Stochastic

The words that we have considered so far have many connotations. Nevertheless we may be comforted by the word "stochastic" introduced by James Bernoulli (1654-1705) explicitly for the purposes of probability theory. Stochastic (adj.) is taken from the Greek word "stochazesthai" (v.) which means to shoot at a mark, and which was used by Plato in the sense of to aim at an ideal. Although mathematicians have been using the word stochastic since the time Bernoulli introduced it, it has not yet entered common language and so lacks such connotations as are associated with the other words examined in this chapter.

Blaise Pascal wrote: Those who are accustomed to judge by feeling do not understand the process of reasoning, for they would understand at first sight, and are not used to seek for principles. And others, on the contrary, who are accustomed to reason from principles, do not at all understand matters of feeling, seeking principles, and being unable to see at a glance.

CHAPTER 3. PROBLEMS OF DIVISION

> Blaise Pascal wrote: Chance gives rise to thoughts, and chance removes them; no art can keep or acquire them.

3.1 Pascal and Fermat

Blaise Pascal and Pierre Fermat are generally credited with the origin of Theory of Probability. The Chevalier de Mere proposed certain questions to Pascal. Starting in 1654, Pascal corresponded with Pierre de Fermat on the subject of these questions. Fermat wrote letters in reply.

One of Pascal's letters discusses the "Problem of Points." Two players each need a given number of points in order to win; if they separate without playing out the game, how should the stakes be divided between them?

The question amounts to asking, at any given stage of the game, what is the probability which each player has of winning the game? In the discussion between Pascal and Fermat, it is supposed that the players have equal chances of winning a single point. Pascal's account is:

> The following is my method for determining the share of each player, when, for example, two players play a game of three points and each player has staked 32 pistoles.

> Suppose that the first player has gained **two points** and the second player **one point**; they have now to play for a point on this condition, that if the first player gains he takes all the money which is at stake, namely 64 pistoles, and if the second player gains each player has two points, so that they are on terms of equality, and if they leave off playing each ought to take 32 pistoles Thus, if the first player gains, 64 pistoles belong to him, and if he loses, 32 pistoles belong to him. If, then, the players do not wish to p ay this game, but to separate without playing it, the first player would say to the second. "I am certain of 32 pistoles even if I lose this game, and as for the other $32 pistoles perhaps I shall have them and perhaps you will have them; the chances are equal. Let us then divide these 32 pistoles equally and give me also the $32 pistoles of which I am

certain." Thus the first player will have **48 pistoles** and the second **16 pistoles**.

Next, suppose that the first player has gained **two points** and the second player **none**, and that they are about to play for a point; the condition then is that if the first player gains this point, he secures the game and takes the 64 pistoles, and if the second player gains this point the players will then be in the situation already examined, in which the first player is entitled to 48 pistoles, and the second to 16 pistoles. Thus if they do not wish to play, the first player would say to the second, 'If I gain the point, I gain 64 pistoles; if I lose it I am entitled to 48 pistoles. Give me then the 48 pistoles of which I am certain, and divide the other 16 equally, since our chances of gaining the point are equal.' Thus the first player will have **56 pistoles** and the second player **8 pistoles**.

Finally, suppose that the first player has gained **one point** and the second player **none**. If they proceed to play for a point the condition is that if the first player gains it the players will be in the situation first examined, in which the first player is entitled to 56 pistoles; if the first player loses the point each player has then a point, and each is entitled to 32 pistoles. Thus if they do not wish to play, the first player would say to the second "Give me the 32 pistoles of which I am certain and divide the remainder of the 56 pistoles equally, that is, divide 24 pistoles equally." Thus the first player will have the sum of 32 and 12 pistoles, that is **44 pistoles**, and consequently the second will have **20 pistoles**.

3.2 The problem of Arthur and Bill.

This section gives an analysis of the foregoing section. The first player is Arthur and the second player is Bill. Each year Arthur and Bill, equally good sailors, compete with each other at Marblehead in a series of 5 sailboat races, the winner of 3 races out of 5 taking the yearly prize of $6400. Nevertheless in three consecutive years, namely 1960, 1961, and 1962, this competition was called off before completion because of hurricane weather. The results of these three years are as follows:

1961: Arthur won 2 races and Bill won 1 race.

1962: Arthur won 2 races and Bill won no race.

1963: Arthur won 1 race and Bill won no race.

How should Arthur and Bill split the prize of $6400 in each of these years?

The following conversation takes place.

Arthur. Let us look at 1961. We sailed 3 races: race 1, race 2, and race 3.

I won 2 of them and you won 1 of them. Suppose we did sail race 4 in 1961. If I won this race, then I would take the prize of $6400. If you won this race, then we would have both won 2 races each, so that we would be on terms of equality and so should each have $3200.

Bill. That is true.

Arthur. So if I won race 4, I would gain $6400. If I lost race 4 then I am entitled to $3200. Hence give me the $3200 which I am certain of, and divide the remaining $3200 equally.

Bill. Thus your share would be $3200 plus $1600 which is **$4800**.

Arthur. Yes. Your share is the remaining **$1600**.

Bill. This is fair, but I would reason this way. In 1961 we sailed 3 races of which you won 2 and I won 1. Certainly the prize of 1961 would be determined by 2 more races, namely race 4 and race 5. Now the possible results of these two races are

AA, AB, BA, BB*

where A stands for you (Arthur) winning a race and B stands for me (Bill) winning a race. Because I had to win both of these races to take the prize, only the last possible result, namely BB is favorable for me. That is why I put an asterisk beside it.

Arthur. So you only had 1 chance out of 4 of taking the prize.

Bill. That is right, so my share in 1961 would be 1/4 of $6400 which is $1600.

Arthur. That agrees with my method.

Bill. This is a contented solution for 1961.

Arthur. Now let us look at 1962. We sailed 2 races: race 1 and race 2. Because I won these 2 races, I would need to win only one more race to take the $6400 prize

Bill. And since I have won no races, I would have to win 3 races to take the prize.

Arthur. Two races were sailed. Suppose we did sail race 3. If I won this race then I would take the entire prize of $6400. On the other hand, if you won this race, then there would be 2 races won by me and 1 race won by you. Hence we would be in the same situation as 1961, namely that I am entitled to $4800 and you to $1600.

Bill. I follow you. If you won race 3, then you deserve $6400. If you lost race 3, then you deserve $4800 and I deserve $1600. But race 3 was not sailed.

Arthur. Yes, so if I won race 3, I would gain $6400. If I lost race 3, I am entitled to $4800 anyhow. Hence give me the $4800 which I am certain of, and divide the remaining $1600 equally.

Bill. Your share then is $4800 plus $800 which is **$5600**.

Arthur. And your share is the remaining **$800**.

Bill. Still I prefer my solution, which looks like this. In 1962 we sailed 2 races, both of which you won. Certainly the prize would be determined by sailing 3 more races, namely race 3, race 4, and race 5. The possible results of these 3 additional races are

AAA	ABA	ABB	BBA
AAB	ABB	BAB	BBB*

The table shows that there are 8 possible results. Now I have put an asterisk beside the only one favorable to me, the one in which I win all the races. All the other possible results are favorable to you.

Arthur. Your method is becoming clear. It shows that you have only 1 chance in 8 of taking the prize, so your share is 1/8 of $6400 which is $800, and my share is 7/8 of $6400 which is $5600.

Bill. That is right. Our two methods give the same answer, so we reached a cheerful solution for 1962.

Arthur. **Let us look at 1963**. We sailed only 1 race. I won this race, and then we did not sail any more races. But suppose we did sail race 2. If I had won race 2, then we would be in the same situation as 1962, in which I am entitled to $5600. On the other hand, if you had won race 2, then we would have won 1 race each, so that we would each be entitled to $3200.

Bill. I follow you. If you won race 2 then you deserve $5600. If you lost race 2 then you deserve $3200. But race 2 was not sailed.

Arthur. Yes, so give me the $3200 which I am certain of and divide the remainder of the $5600 equally. The remainder is $5600 minus $3200 which is $2400, so $2400 should be divided equally.

Bill. Thus your share for 1963 is $3200 plus $1200, which is **$4400**.

Arthur. And your share would be $800 plus half of the $2400 that is divided equally. That is, your share for 1963 is $800 plus $1200, which is **$2000**.

Bill. Now this is my method. In 1963 one race was sailed which you won. Now the prize would be determined by sailing 4 more races. The possible results of these 4 additional races are

AAAA	ABAA	BAAA	BBAA
AAAB	ABAB	BAAB	BBAB
AABA	ABBA	BABA	BBBA
AABB	ABBB*	BABB*	BBBB*

Notice that I have put an asterisk beside the ones favorable to me. As you see there are 16 possible results, and 5 asterisks, so my chances are 5 out of 16. Hence I should have 5/16 of the $6400 which is $2000 and you should have the remaining $4400.

Arthur. Everything is settled. We have found an acceptable solution for each year.

3.3 The problem of Ann and Betty

Ann and Betty, each an equally good player, enter a contest where a prize of $64 is offered. The prize is to be awarded to the one who wins 2 out of 3 games. They play one game and Ann wins, but then the contest is called off. How should the prize be divided between Ann and Betty?

The following conversation takes place.

Ann. I think that the prize should be divided proportionally to the number of games already won by each player. Since there has been only

1 game played, and I won it, it seems to me that I should have the entire prize of $64.

Betty. That division of the prize is unfair. If we did play the other 2 games, I would have a chance of winning both of them, and thereby win the $64. Remember the prize was to be awarded not to the person who wins the most games in an unspecified number of games, but to the person who first wins 2 games.

Ann. All that seems reasonable, but still I don't think we can do anything about it. The fact remains that I won the only game played.

Betty. Yes, but we must also look at the games not played. Now what are your chances of taking the prize of $64? As I see it, you have 2 chances of taking the prize. One chance is that you win game 2. The other chance is that you lose game 2 but win game 3. On the other hand, I have 1 chance of taking the prize, namely by winning both the remaining games; that is, by winning game 2 and game 3.

Ann. I see. Since I have 2 chances out of 3 of taking the prize, I should have 2/3 of $64 or $42.67 and you should have 1/3 of $64, or $21.33

Betty. Yes, that is good, $42.67 for you, and $21.33 for me.

Ann. But still, I think I deserve more. Suppose we reason this way. If the contest were not called off then we would play game 2. If I win game 2 then I would take the entire $64. On the other hand if you win game 2, then we would have each won 1 game. In this case, we would be on terms of equality and thus the prize should be shared equally, that is $32 each. In other words, if I win game 2 then $64 belongs to me, whereas if I lose game 2 then $32 belongs to you.

Betty. I am not sure that I understand.

Ann. Look at it this way. I am certain of $32 even if I lose game 2. As for the other $32 perhaps I will have it and perhaps you will have it, as the chances are equal. Let us then divide this $32 equally and also give me the $32 which I am certain of.

Betty. Now I think I understand. The prize should be divided so that you receive $16 plus $32 or $48 and I receive the remaining $16.

Ann. Yes, that's what I mean. You should have $16 of the $64.

Betty. And you should have $48 of the $64. But I am still confused why my solution which gave you only $42.67 is not right.

Ann. Let A stand for me, Ann, winning and let B stand for you, Betty, winning. We are sure that the prize will be decided by game 2 and game 3. Let us look at the possible results of these 2 games. They are

AA   AB   BA   BB

where, for example, AB stands for me winning game 2 and you winning game 3.

Betty. I see. Then AA stands for you winning both games and BB stands for me winning both games. Also AB stands for you winning game 2 and for me winning game 3, and BA for the opposite.

Ann. Yes. Now there are 4 possible results in total. Let me put an asterisk beside any one for which you take the prize. Thus I will put an asterisk beside the BB so we have

AA   AB   BA   BB*

Betty. Let me see. Because you have won 1 game, namely game 1, you only need to win one more game to collect the prize.

Ann. And since you have won no game you need to win 2 games to collect the prize.

Betty. I see. I can only take the prize by winning both game 2 and 3. That is, the only possible result of the remaining games for which I collect the prize is B3.

Ann. Yes, that's why I put the asterisk beside BB. Now, as you see, there is only 1 asterisk out of the 4 possible results.

Betty. So it means that I have only 1 chance in 4 of collecting the prize.

Ann. Yes, so you should have 1/4 of the prize, which is $16.

Betty. And because you have 3 chances out of 4 you should have 3/4 of the prize, which is $48.

Ann. That is right.

Betty. But still there s one thing that bothers me. In your method, it is **supposed** that 2 more games (that is, game 2 and game 3) will be played. But this is not necessarily the case, for it is possible that you may win game 2, so that the prize will be determined after only 2 games in total (that is, game 1 and game 2).

Ann. Of course, that is true.

Betty. Then there are only 3 possible results, namely

A   BA   BB*

Now I have put an asterisk beside BB, which is the one favorable to me. You see that now I have 1 chance out of 3 so that I should have 1/3 of $64 or $21.33, instead of only $16 as you suggest.

Ann. Let me give the following answer to your objection. Although it is possible that the prize may be determined in 2 games, we are at liberty to conceive that we agree to play 3 games beforehand. Even if the prize is determined in 2 games, the superfluous game 3 will make no difference in the decision.

Betty. In other words, you have both

AA   AB

as 2 different possible results in your scheme. In my scheme, corresponding to your AA and AB, I have only the possible result

A

The reason for me just using A is that once you win game 2, the prize is determined for you, and there is no need to play game 3.

Ann. Yes, but certainly we could conceive that we play game 3, so that instead of the possible result A in your scheme we should have the 2 possible results AA and AB that are in my scheme. So your share really is $16.

Betty. Well, anyway, $16 is better than nothing, so I'll be happy to take that.

Ann. Yes, that is a comfortable solution.

3.4 Analysis of 4 different situations.

Suppose that a contest for a prize of $64 between A and B is called off before all the games are played. Both A and B are equally good players.

Each game is won by either A or B, as the possibility of a tied game is ruled out.

SITUATION 1. Suppose that A lacks 1 game in order to win the contest, and that B also lacks 1 game in order to win the contest. If 1 more game is played, then there are 2 possible results, namely, either A wins it or B wins it, as shown pictorially by the tree diagram in Figure 1.

Game 1

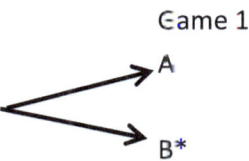

Figure 1. Possible results

We have placed an asterisk beside the possible result favorable to B winning the contest. Since there are 2 possible results, and only 1 asterisk, it follows that B has 1 chance in 2 of winning the contest. Thus the probability of B winning the contest is 1/2, so his expectation is

1/2 of $64 = $32

Thus B's fair share of the $64 prize is $32, and it follows that A's fair share is $64 - $32 = $32 .

SITUATION 2. Suppose that A lacks 1 game in order to win the contest, and B lacks 2 games in order to win the contest. If 1 more game (called game 1) is played, then there are 2 possible results, namely either A wins it or B wins it. If A wins it, then A wins the contest. On the other hand, if B wins it, then another game (called game 2) must be played. Thus we have the 3 possible results as shown by the tree diagram in Figure 2. In symbols, the 3 possible results are

A   BA   BB*

We have placed an asterisk beside the possible result favorable to B.

Game 1   Game 2

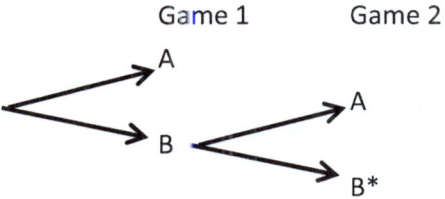

Figure 2. Possible results (with unequal ranges)

However these 3 possible results do not all have equal ranges, so they do not represent equal chances. Therefore, if A wins game 1, we must suppose that game 2 is played anyhow (even though the issue of the prize is no longer at stake). Thus we have 2 times 2 = 4 possible results as shown by the tree diagram in Figure 3.

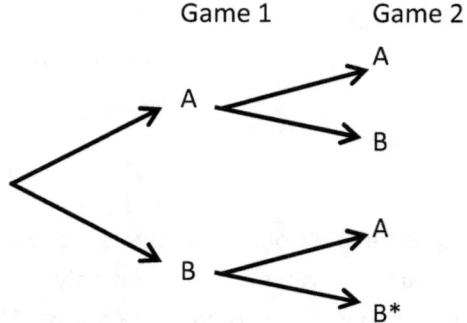

Figure 3. Possible results (with equal ranges)

The 4 paths in this tree diagram have equal length, namely 2 games. Each of these 4 possible results is an equal chance. Thus B has 1 chance (indicated by the asterisk) in 4 to win the prize of $64. Thus his probability is 1/4, and his expectation is 1/4 of $64 = $16. The expectation of A is the remaining $48.

SITUATION 3. Suppose that A lacks 1 game, and that B lacks 3 games, to win the contest.

By playing 1 more game (called game 1), and (if need be) another game (called game 2), and again (if need be) another game (called game 3), the issue of the prize is determined. The 4 possible results are shown by the tree diagram in the Figure 4.

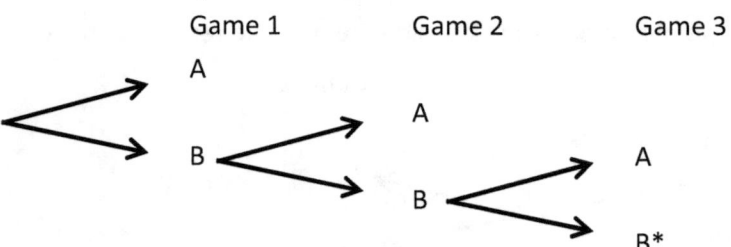

Figure 4. Possible results (with unequal ranges)

We have placed an asterisk beside the possible result favorable to B. Nevertheless these 4 possible results do not have equal ranges, so they do not represent equal chances. Therefore we must suppose that 3 games are played anyhow (whether they are needed or not to determine the prize). Thus we have 2 times 2 times 2 = 8 possible results as shown by the tree diagram in Figure 5.

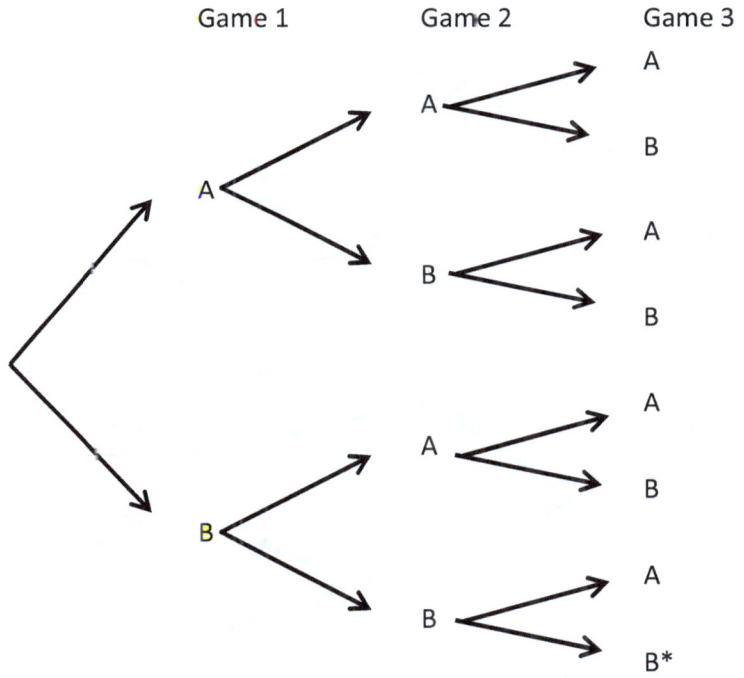

Figure 5. Results (with equal ranges)

The 8 paths in this tree diagram have equal length, namely 3 games. Each of these 8 possible results is an equal chance. Only 1 (indicated by the asterisk) is favorable to B winning the contest. Hence B has 1 chance in 8 to win the $64. Thus his probability of winning the $64 is 1/8 and his expectation is 1/8 times $64 = $8.

On the other hand, A has 7 chances in 8 of winning the $64. Thus his probability of winning the $64 is 7/8 and his expectation is 7/8 times $64 = $56.

SITUATION 4. Suppose that A lacks 2 games, and that B lacks 3 games, to win the contest.

The contest is determined by at most 2 + 3 - 1 = 4 more games, called games 1, 2, 3, and 4. The 10 possible results are shown in the Figure 6, where an asterisk is placed beside each one favorable to B.

Game 1	Game 2	Game 3	Game 3

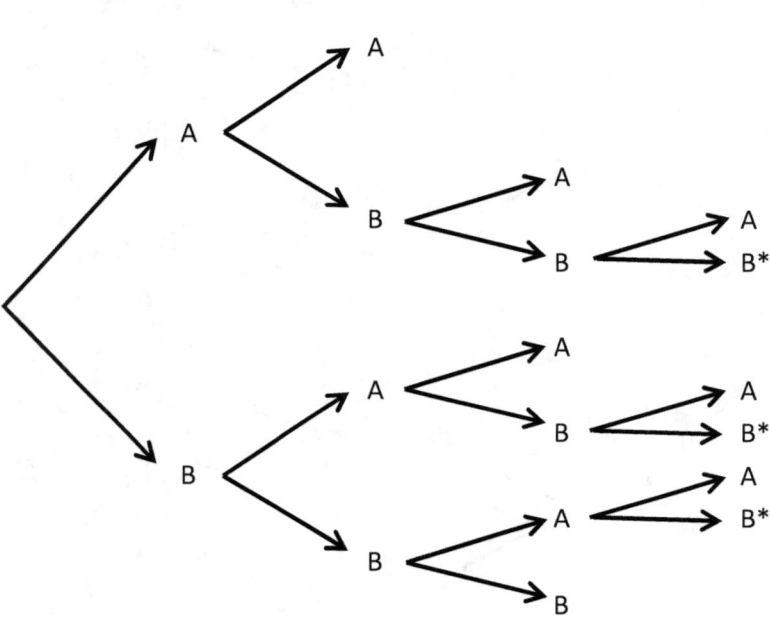

Figure 6. Possible results (with unequal ranges)

As the tree diagram shows, these 10 possible results do not have equal ranges, and hence they do not each represent the same chance. Thus we suppose that all of the 4 games are played, even when the issue of the contest is no longer in question. We thereby obtain

$2 \times 2 \times 2 \times 2 = 16$

possible results shown by the tree diagram in Figure 7.

The 16 paths in the tree diagram have equal length, namely 4 games. We have placed an asterisk beside each of the 5 possible results favorable to B. Thus B has 5 chances in 16, or a probability of 5/16, to win the prize of $64, so his expectation is 5/16 times $64 = $20. The remaining 11 chances are favorable to A. Thus A has 11 chances in 16, or a probability of 11/16, to win the $64, so his expectation is 11/16 times $64 = $44.

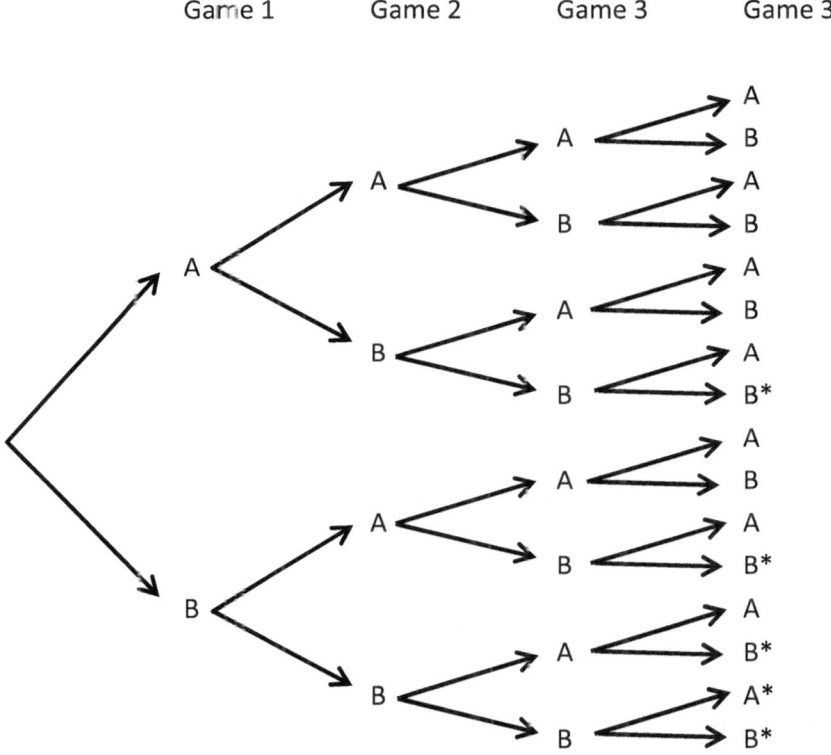

Figure 7. Possible results (with equal ranges)

3.5 Assignment of probabilities to simple events

Up to now we have assumed that A and B are equally good players, so the chances are 50:50 for either winning a single game. In other words, A has 50 chances in 100 to win a single game, whereas B has the other 50 chances in 100 to win the game, as we have ruled out the possibility of a tie.

Instead of saying that A has 50 chances in 100 to win a single game, we may alternatively say that A has the probability

p = 50/100 = 1/2 = 0.5

to win a single game. Instead of saying that B has the other 50 chances in 100 to lose a single game, we may alternatively say that B has the probability

q = 50/100 = 1/2 = 0.5

to lose a single game.

Now suppose that A is a better player than B. For example, we may suppose that the chances are 60:40 for winning a single game. In other words A has 60 chances in 100 to win a single game, whereas B has the remaining 40 chances in 100 to win the game. Then the probability that A wins a single game is

p = 60/100 = 6/10 = 0.6

and the probability that B wins the single game is

q = 40/100 = 4/10 = 0.4

Because we have excluded the possibility of a tie game, we see that the probability p plus the probability q always adds up to 1; that is, p + q = 1.

We may now assign probabilities to the various events in the 4 situations considered in the foregoing section.

SITUATION 1. A lacks 1 game and B lacks 1 game. Hence at most 1 + 1 − 1 = 1 more game must be played. We suppose this game is played. Then there are 2 possible results, namely A, B. Each of these possible results makes up a simple event, and we assign probabilities to each simple event as follows:

Simple event	Probability
A	p
B	q

The sum of the probabilities of the 2 simple events is p + q = 1. This sum is 1 because we have ruled out the possibility of a tie. This sum being 1 insures us that we have not forgotten any simple event.

SITUATION 2. A lacks 1 game and B lacks 2 games.

Hence at most 1 + 2 − 1 = 2 more games must be played. We suppose that 2 games are played. Then there are 2 times 2 = 4 possible results, namely AA, AB, BA, BB. Each possible result makes up a simple event. To the simple event of 2 A's in a row, namely AA, we assign probability p p

$= p^2$. To the simple event of A followed by B namely A B we assign probability $p\,q$, To the simple event of B followed by A, namely BA , we assign probability $q\,p$. Because $p\,q = q\,p$, we see that the simple event A B has the same probability as the simple event B A, namely probability p q. Similarly, we assign probability q^2 to the simple event BB. We have

Simple event	Probability
AA	p^2
AB	$p\,q$
BA	$q\,p$
BB	q^2

The sum of the probabilities assigned to the 4 simple events is

$$p^2 + p\,q + q\,p + q^2 = p^2 + 2\,p\,q + q^2$$

But we can factor this expression into a perfect square. Thus this sum is

$$p^2 + 2\,p\,q + q^2 = (p + q)^2 = 1^2 = 1$$

This sum being 1 serves as a check to tell us that we have included all the simple events.

SITUATION 3. A lacks 1 game and B lacks 3 games.

Hence at most $1 + 3 - 1 = 3$ more games must be played. We suppose that these 3 games are played, so there are 2 times 2 times 2 = 8 possible results. Each possible result makes up a simple event. To the simple event AAA we assign probability $p\,p\,p = p^3$. To the simple event AAB we assign probability $p\,p\,q = p^2\,q$. To the simple event ABA we assign probability $p\,q\,p = p^2\,q$. In other words, we assign the factor p for each A and the factor q for each B in any simple event. Thus we have

Simple event	Probability
AAA	p^3
AAB	$p^2\,q$
ABA	$p^2\,q$
ABB	$p\,q^2$
BAA	$p^2\,q$
BAB	$p\,q^2$
BBA	$p\,q^2$
BBB	q^3

Now:

1 simple event	probability p^3
3 simple events	probability $p^2 q$
3 simple events	probability $p q^2$
1 simple event	probability q^3

Thus the sum of the probabilities of the simple events is

$$p^3 + 3 p^2 q + 3 p q^2 + q^3$$

which can be factored into a perfect cube, that is, $(p + q)^3 = 1^3 = 1$.
Because the sum is 1, we are assured that we have not forgotten any simple events.

SITUATION 4. A lacks 2 games and B lacks 3 games.

Hence at most $2 + 3 - 1 = 4$ more games must be played. We suppose that these 4 games are played, so that there are 2 times 2 times 2 times 2 = 16 possible results. Each possible result makes up a simple event. We assign the probabilities to the simple events as follows:

Event	Prob.	Event	Prob.	Event	Prob.	Event	Prob.
AAAA	p^4	ABAA	$p^3 q$	BAAA	$p^3 q$	BBAA	$p^2 q^2$
AAAB	$p^3 q$	ABAB	$p^2 q^2$	BAAB	$p^2 q^2$	BBAB	$p q^3$
AABA	$p^3 q$	ABBA	$p^2 q^2$	BABA	$p^2 q^2$	BBBA	$p q^3$
AABB	$p^2 q^2$	ABBB	$p q^3$	BABB	$p q^3$	BBBB	q^4

Now:

1 simple event	probability p^4
4 simple events	probability $p^3 q$
6 simple events	probability $p^2 q^2$
4 simple events	probability $p q^3$
1 simple event	probability q^4

Thus the sum of the probabilities of the simple events is

$$p^4 + 4 p^3 q + 10 p^2 q^2 + 4 p q^3 + q^4$$

This expression can be factored into a perfect quadratic; that is,

$$(p + q)^4 = 1^4 = 1$$

Because the sum is 1, we have a check that our list includes all the simple events for this situation.

3.6 Probabilities of compound events.

Once we know the probabilities of all the simple events, we can easily find the probabilities of compound events. Whereas a simple event is made up of 1 possible result, a compound event is made up of 2 or more possible results

Let us now see how we can find the probability of A winning the contest and the probability of B winning the contest for each of the 4 situations described in the foregoing 2 sections. We shall always put an asterisk next to any possible result favorable to B.

SITUATION 1. A lacks 1 game and B lacks 1 game. From the foregoing section we have the table:

Simple event	Probability
A	p = probability of A winning the contest
B*	q = probability of B winning the contest

In the above table we have indicated that the probability of A winning the contest is p and the probability of B winning the contest is q. In this situation, of course, the event of A winning the contest is a simple event, and likewise the event of B winning the contest is a simple event. For example, if A and B are equally good players, the respective probabilities of their winning the contest is p = 1/2, q = 1/2.

SITUATION 2. A lacks 1 game and B lacks 2 games. From the foregoing section, we have the table:

Simple event	Prob.	Compound event	Prob.	Description of probability
AA	pp			probability of A winning the contest
AB	pq	AA, AB, BA	$pp + pq + qp$	
BA	qp			
BB	qq	BB	qq	probability of B winning the contest

Thus the probability that A wins the contest is $p^2 + 2pq$ and the probability that B wins the contest is q^2. In this situation, the event of A winning the contest is a compound event, whereas the event of B winning the contest is a simple event. For example, let us suppose that A and B are equally good players, so that p and q are each equal to 1/2. Then the probability of A winning the contest is

$$p^2 + 2pq = (1/2)(1/2) + 2(1/2)(1/2) = (1 + 2)/4 = 3/4$$

Here $1 + 2 = 3$ is the number of favorable chances, and 4 is the total number of chances. The sum of these 2 probabilities is $3/4 + 1/4 = 1$, as we expect.

SITUATION 3. A lacks 1 game and B lacks 3 games.

The following table of probabilities is arranged from the corresponding table in the foregoing section. This rearrangement was made so that all simple events with the same probability would be adjacent. The rearranged table is

Simple event	Prob.	Compound event	Prob.	Description of probability
AAA	p^3	AAA	p^3	$p^3 + 3p^2q + 3pq^2$
AAB	p^2q			=probability of A
ABA	p^2q	AAB, ABA,	$3p^2q$	winning the contest
BAA	p^2q	BAA		
ABB	pq^2			
BAB	pq^2	ABB, BAB,	$3pq^2$	
BBA	pq^2	BBA		
BBB*	q^3	BBB*	q^3	q^3= probability of B winning the contest

Thus the probability that A wins the contest is

$$p + 3p^2q + 3pq^2$$

whereas the probability that B wins the contest is q^3. In this situation, the event of A winning the contest is a compound event, and the event of B winning the contest is a simple event. For example, if A and B are equally good players, then p and q are each equal to 1/2. The probability of A winning the contest is

$(1 + 3 + 3)/8 = 7/8.$

where 1+3+3=7 is the number of favorable chances and 8 is the number of all chances. Likewise the probability that B wins the contest is

$q^3 = 1/8$

where 1 is the number of favorable chances and 8 is the number of all chances. The sum of these 2 probabilities is 7/8 + 1/8 = 1, as we expect.

SITUATION 4. A lacks 2 games and B lacks 3 games.

The table (rearranged so that simple events with the same probability are adjacent) is

Simple event	Prob.	Compound event	Prob.	Description of probability
AAAA	p^4	AAAA	p^4	
AAAB	$p^3 q$			$p^4 + 4 p^3 q + 6 p^2 q^2$
AABA	$p^3 q$	AAAB, AABA,	$4 p^3 q$	=probability of A
ABAA	$p^3 q$	ABAA, BAAA		winning the contest
BAAA	$p^3 q$			
AABB	$p^2 q^2$			
ABAB	$p^2 q^2$	AABB, ABAB,		
ABBA	$p^2 q^2$	ABBA, BAAB	$6 p^2 q^2$	
BAAB	$p^2 q^2$	BABA, BBAA		
BABA	$p^2 q^2$			
BBAA	$p^2 q^2$			
ABBB*	$p q^3$			$4 pq^3 + q^4$ = probability
BABB*	$p q^3$	ABBB, BABB,		of B winning
BBAB*	$p q^3$	BABB, BBBA	$4 pq^3$	the contest
BBBA*	$p q^3$			
BBBB*	q^4	BBBB	q^4	

The probability that A wins the contest is

$p^4 + 4 p^3 q + 6 p^2 q^2$

and the probability that B wins the contest is

$4 p q^3 + q^4$

In this situation, both the events of A winning the contest and the event of B winning the contest are compound events. For example, if p = 1/2, q = 1/2, then the probability of A winning the contest is (1 + 4 + 6)/16 = 11/16 and the probability of B winning the contest is (4 + 1)/16 = 5/16.

In other words, A has 11 chances in 16, and B has 5 chances in 16, to win the contest. The sum of the probability of A winning the contest plus the probability of B winning the contest is

11/16 + 5/16 = 16/16 = 1

as we expect.

3.7 Some problems and solutions

Problem 1. Suppose that A and B are equally good players and that A lacks 2 games and B lacks 2 games to win a contest for a prize of %64 . What are their expectations?

Solution. The contest will be decided in at most 2 + 2 - 1 = 3 more games, which we suppose are played. The 8 possible results are

AAA	BAA
AAB	BAB*
ABA	BBA*
ABB*	BBB*

where we have placed an asterisk beside the ones favorable to B winning the contest. Since each possible result is 1 chance, there are 4 chances in 8 favorable to A winning the contest, and 4 chances in 8 favorable to B winning the contest. Hence each has a probability of winning the contest equal to 4/8 = 1/2 so each has an expectation of

(1/2) ($64) =$32.

Problem 2. For each game A has 2 chances in 3 of winning, whereas B has 1 chance in 3 of winning. In a contest for $63, A lacks 1 game and B lacks 2 games. What are their expectations?

Solution 1. The contest will be decided in at most 1+2-1= 2 games, both of which we suppose are played. We have the table, in which we have put an asterisk beside the possible result favorable to B.

Simple event	Prob.	Compound event	Prob.
AA	p^2	AA, AB, BA	$p^2 + 2\,pq$ =probability of A winning the contest
AB	$p\,q$		
BA	$q\,p$		
BB*	q^2	BB*	q^2 = probability of B winning the contest

The probability that A wins the contest is

$$p^2 + 2\,pq$$

The probability that B wins the contest is q. Since A has 2 chances in 3 for winning each single game, we set p = 2/3 and since B has 1 chance in 3 for winning each single game, we set q = 1/3. Thus the probability that A wins the contest is

$$p^2 + 2\,pq = (2/3)(2/3) + 2(2/3)(1/3) = (4+4)/9 = 8/9$$

The probability that B wins the contest is

$$q^2 = (1/3)(1/3) = 1/9$$

Hence the expectation of A is 8/9 of $63 = $56 and the expectation of B is 1/9 of $63 = $7.

Solution 2. Because A has 2 chances in 3 and B has 1 chance in 3, to win each game with each other, let us represent the possible results of each game by the 3-way fork:

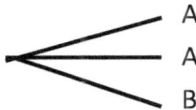

By using this 3-way fork, each of the 3 possible results of a single game represents 1 chance. Now the contest will be decided in at most 2 more games, which we assume are played.

There are 9 chances in all, namely

AA	AA	BA
AA	AA	BA
AB	AB	BB*

The possible results are shown by the tree-diagram in Figure 8.

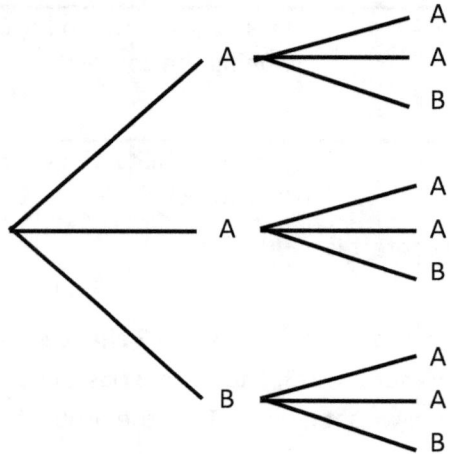

Figure 8. Possible results as equal chances.

Only one of these (i.e. the one with the asterisk) is favorable to B winning the contest. Thus the expectation of B is

1/9 of $63=$7

The expectation of A is the remaining $56.

Problem 3. Suppose that A has 2 chances in 3, and B has 1 chance in 3, to win each game between them. They agree to compete in a contest made up of 3 games. The prize of $54 is awarded to A if he wins at least 2 games out of the 3 games. Otherwise, the prize is awarded to B. What are their expectations?

Solution. There are 8 possible results for the 3 games. Thus we have the table:

Simple event	Probability	
AAA	p^3	probability of A
AAB	p^2q	winning the
ABA	p^2q	contest
BAA	p^2q	
ABB*	pq^2	probability of B
BAB*	pq^2	winning the
BBA*	pq^2	contest
BBB *	q^3	

Here we have put an asterisk beside each possible result favorable to B winning the contest. The probability of A winning the contest is

$$p^3 + 3\,p^2q$$

We set p = 2/3 and q = 1/3 so the probability of A winning the contest is

$(2/3)^3 + 3(2/3)^2(1/3) = (8 + 12)/27 = 20/27$

so the expectation of A is (20/27)($54) = $40. Likewise the expectation of B is (7/27) ($54) = $14. Thus, although A has 2 chances in 3, that is, 18 chances in 27, to win each single game, he has 20 chances in 27 to w n the contest.

Problem 4. In any single game, one of A, B, C must win, and they are all equally good players. Suppose a lacks 1 game, B lacks 1 game, and C lacks 2 games, to win a contest. What are their respective chances of winning the contest?

Solution. The contest will be decided in at most 2 games. The 3 times 3 = 9 possible results are

AA	BA*	CA
AB	BB*	CB*
AC	BC*	CC**

We have placed an asterisk beside those favorable to B, and 2 asterisks beside that favorable to C . Each possible result is 1 chance. Thus A has 4 chances in 9, B has 4 chances in 9 , and C has 1 chance in 9 to win the contest.

Problem 5. Let the probability that A wins a single game be p and the probability that B wins a single game be q. We assume that a game cannot end in a draw, so p + q = 1. Suppose that A must win 1 more game, and B must win 2 more games, to win a contest. What is the probability that A wins the contest?

Solution 1. To win the contest, A can win by either

(1)   By playing exactly 1 more game

(2)   or by playing exactly 2 more games

These 2 events are incompatible, that is, they cannot both happen together. To win the contest by playing exactly 1 more game, A must win that game. His probability of doing this is p. To win the contest by playing exactly 2 more games, A must lose game 1 and win game 2. His probability of doing this is qp. Thus the probability that wins the contest is p + q p.

Solution 2. The possible results and the probabilities of each branch of the tree are shown by the tree diagram in Figure 9.

Game 1   Game 2

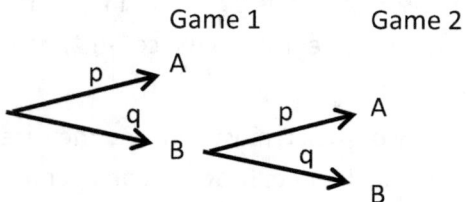

Figure 9. Possible results

Hence the simple events and their probabilities are

Simple event	Prob.	Compound event	Probability
AA	p	AA, BA	p2+qp=probability of A winning the contest
BA	qp		
BB*	q2	BB*	q2= probability of B winning the contest

The possible result A means that A wins by playing exactly 1 more game
The possible result BA means that A wins by playing exactly 2 more
games. Thus the probability of A winning the contest is p+ qp. We have
placed an asterisk beside the possible result favorable to B winning the
contest. His probability of winning the contest is thus q^2.

Solution 3. The contest is decided in at most 2 more games. We may
assume that both of these games are played. The possible results are
shown by the tree diagram in Figure 10.

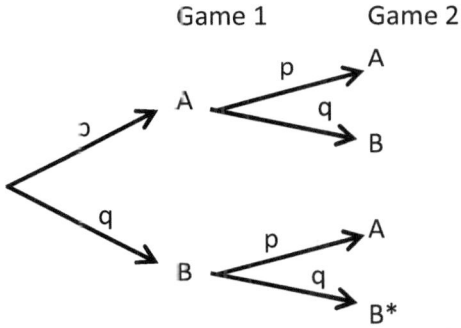

Game 1   Game 2

Figure 10. Possible results

The winning by player A entails the compound event AA, AB, Ba. Hence
the probability of A winning the contest is $p^2 + p\,q + q\,p = p(p + q) + qp$.
Because $p + q = 1$, this probability is $p + qp$, which checks with the
answer given in Solutions 1 and 2 .

> Blaise Pascal wrote: When a natural discourse paints a passion or an
> effect, one feels within oneself the truth of what one reads, which
> was there before, although one did not know it. Hence one is
> inclined to love him who makes us feel it, for he has not shown us his
> own riches, but ours. And thus this benefit renders him pleasing to
> us, besides that such community of intellect as we have with him
> necessarily inclines the heart to love.

CHAPTER 4. PROBABILITY MODELS

Blaise Pascal wrote: We are generally more effectually persuaded by reasons we have ourselves discovered than by those which have occurred to others.

4.1 Models

One of the purposes of scientific investigation is the construction of theoretical models of real-life (or real-world) situations. A model is conceptual in that it represents a mental image of the underlying situation. All scientific models must be ultimately derived from observations. Every model is intrinsically "open-ended"; that is, it is not perfect but is subject to future improvement.

To attain good models both the scientist and the mathematician must work closely together. The science in which mathematical models have been the most fruitful is physics. A well-known example of a mathematical model is Newton's second law. It states that force F, mass m, and acceleration a are related by the formula F= ma. This formula represents the mathematical model of the law of nature known as Newton's second law.

4.2 Sure Situations

Any real-life (or real-world) situation represents a relative combination of circumstances. A large part of mathematics is concerned with the building of mathematical models of sure situations. A sure situation is one in which there is no degree of uncertainty; it has all the precision and predictability of a mathematical equation.

An example of a sure situation is represented by Newton's second law, just discussed above. It involves 3 quantities; namely, force F, mass m, or acceleration a. Whenever any 2 of the 3 are given, the third quantity may be found by solving the mathematical equation F= ma.

A sure situation is identified by

(1) its circumstances

(2) its result

Its circumstances are the antecedent conditions. Whenever the circumstances-are fulfilled, the result occurs. For example, from Newton's second law we know that whenever a force is applied to a mass (the circumstances), the mass accelerates (the result).

Sure situations are very common, so let us list a few more examples:

1. Circumstances: Water is cooled to a temperature of 0°C at normal atmospheric pressure. Result: The water freezes.

2. Circumstances: Water is heated to a temperature of 100°C at normal atmospheric pressure. Result: The water boils.

3. Circumstances: Bar of iron is put into water. Result: The bar sinks.

4. Circumstances: An acid is mixed with a base. Result: A salt and water is formed until chemical equilibrium is reached.

5. Circumstances: Two parents, each with a pair of dominant genes, mate. The offspring gets one gene from each parent. Result: The offspring has a dominant pair of genes.

4.3 Unsure situations

The branch of mathematics known as probability theory is concerned with the building of mathematical models of unsure situations. These models are called probability models.

An unsure situation is, one in which there is a degree of uncertainty; it does not have the precision and predictability of a mathematical equation. An unsure situation is identified by

(1) its circumstances

(2) its possible results

Its circumstances, are the antecedent conditions. We recall that a sure situation has only one result. In contrast, an unsure situation has many (that is, more than one) possible results. Whenever the circumstances of an unsure situation are fulfilled, one and only one of the possible results happens.

Here and always we use the word to happen with its English language meaning

to happen: to occur by chance.

The natural result of an unsure situation is therefore a happening.

Unsure situations are very common in human experience. Here are a few examples of stochastic phenomena.

1. Circumstances: A coin is, tossed. Possible results: Either heads or tails.

2. Circumstances: A die is tossed. Possible results: Any one of 1, 2, 3, 4, 5, 6.

3. Circumstances: A card is drawn from the top of a pack of 52 cards. Possible results: A card of any suite and any value.

4. Circumstances: A rain drop falls on a piece of paper. Possible results: Any one of the points on the paper.

5. Circumstances: A seed falls from a tree and lands in a meadow. Possible results: Any one of the points in the meadow.

6. Circumstances: An arrow is shot at a target. Possible results: Any one of the points on the target.

For the unsure situation in question, we must agree on what is meant by the possible results. Thus in Example 1 above, the coin could stand on its edge and thereby not fall heads or tails. Nevertheless it is expedient in most applications to regard heads and tails as the only possible results. Such idealizations are standard practice in scientific model building, and indeed must be made in order to make progress with most problems.

Another idealization is illustrated by Example 4. A rain drop certainly is not a point, but it is expedient to consider it as such in this problem. In Example 6 we have idealized the problem so that the target is large enough or the marksman's aim is good enough so that the arrow always hits the target.

In reference to Example 3, an unstated assumption is that the pack of cards has been well shuffled. On the other hand, if the pack of cards is, ordered so that the ace of spades is on top, and if this fact is known, then the drawing of a card from the top of the pack would not be an unsure situation, but would be a sure situation. Such unstated assumptions as the proper shuffling of cards are usually made in problems connected with probability theory.

4.4 The possible results (or cases) of a stochastic phenomenon.

First of all we need a shorter, and easier to say, expression, for possible result. The word that we will use instead of the expression possible result is case. The use of the word case for this purpose is traditional; and is traceable to the origin of the word in the Latin "cadera" meaning "to fall, to happen".

When a coin falls, there are two cases, namely heads and. tails. When a die falls, there are s x cases, namely 1, 2, 3, 4, 5, 6.

Let us now look more closely into what we mean by a possible result, or case.

We recall that, whenever the circumstances of an unsure situation are fulfilled, one and only one case happens. The phrase "one and only one case happens" means the following three things:

(1) Less than one case cannot happen; that is, the cases are elementa .

(2) One case must happen; that is, the cases are exhaustive

(3) More than one case cannot happen; that is, the cases are exclusive

We want to remember these three things, namely the possible results, or cases, of an unsure situation are EEE:

(1) E for E emental. In saying that the cases are elemental, we mean that the cases are elemental in that they are indecomposable. The individual cases cannot be subdivided or cut up in any way.

(2) E for Exhaustive. In saying that the cases are exhaustive, we mean that the cases exhaust all the possibilities. No cases have been left out.

(3) E for Exclusive. In saying that the cases are exclusive we mean that the happening of one case excludes the happening of the other cases. Two cases cannot happen in any given instance of the situation.

For example, the cases in tossing a die are the faces

1, 2, 3, 4, 5, 6

The faces are EEE. Let us explain why.

(1) E for Elemental: We do not allow the faces of the die to be subdivided in such a way that only part of a face shows. That is, a part of one face is not admitted.

(2) E for Exhaustive: We do not allow the die to stand on an edge, so that no face shows That is, one face must show.

(3) E for Exclusive: We do not allow the die to show 2 faces, say face 4 and face 5, at the same time. That is only one face must show.

4.5 The analysis of cases

When we are confronted with an unsure situation, our first task is to make an analysis of its possible results, or cases. Generally the analysis can be done in several different ways, and the particular way chosen depends upon the problem at hand.

For example, suppose two coins are tossed. Coin 1 has side heads H and side tails T. Coin 2 has side heads h and side tails t. Let us ask what are the possible results? (Or, equivalently, what are the cases?) Suppose there are 3 problems that we are interested in:

Problem 1. Which side appears for each coin?

Problem 2. How many heads appear?

Problem 3. Whether the two coins match (that is, whether both are heads or both are tails) or not match (that is, heads on one coin and tails on the other coin)?

The four cases in Problem 1 are

Hh	(when Coin 1 lands H and that Coin 2 lands h)
Ht	(when Coin 1 lands H and that Coin 2 lands t)
Th	(when Coin 1 lands T and that Coin 2 lands h)
Tt	(when Coin 1 lands T and that Coin 2 lands t)

The three cases in Problem 2 are:

2 heads	(when Hh occurs),
1 heads	(when either Ht or Th occurs),
0 heads	(when Tt occurs)

The two cases in Problem 3 are

Match   (when either 2 heads or 0 heads occurs)

Not match   (when 1 heads occurs)

Each set of cases for the three problems is an admissible set, that is one and only one case happens. However it is seen that the cases for Problem 1 represent a finer analysis than the cases for Problem 2. In other words: If we know which case in the listing for Problem 1 happens, than we may tell which case happens in the listing for Problem 2. Nevertheless, the reverse is not true. Also we see that the analysis for Problem 2 s finer than the analysis for Problem 3.

Next let us consider the situation described in the following problem.

There are 2 urns. Urn 1 contains two white balls and one black ball. Urn 2 contains one white ball and two black balls. An urn is selected at random and from t two balls in succession are drawn at random.

Let us ask the question. What are the possible results? (Or, equivalently, what are the cases?) First let us make our analysis as far as the colors of the balls drawn are concerned. We then have the cases listed in Table 1, where we let W stand for white ball and B stand for black ball.

Case	Urn	Draw 1	Draw 2
1	1	W	W
2	1	W	B
3	1	B	W
4	2	W	B
5	2	B	W
6	2	B	B

Table 1

The analysis shown in Table 1 may be sufficient for many purposes. Nevertheless we may carry out a finer analysis, in which we distinguish between the balls of the same color in an urn. Such an analysis is shown in Table 2 where W1 stands for white ball no. 1, W2 stands for white ball no. 2, etc.

For some purposes the finer analysis of cases shown in Table 2 may be necessary, whereas for other purposes the grosser analysis of cases shown in Table 1 may be adequate. It is important to realize that the cases in a given problem may be analyzed in many different ways, from a very rough analysis to a highly refined one.

Case	Urn	Draw 1	Draw 2
1	1	W1	W2
2	1	W1	B
3	1	W2	W1
4	1	W2	B
5	1	B	W1
6	1	B	W2
7	2	W	B1
8	2	W	B2
9	2	B1	W
10	2	B1	B2
11	2	B2	W
12	2	B2	B1

Table 2

The two conditions that must be met for any analysis of cases are:

Condition 1. Under the circumstances of the unsure situation, one and only one of the cases happens. In other words, the cases are EEE, where the first E stands for elemental, the second E stands for exhaustive, and the third E stands for exclusive.

Condition 2. The analysis is fine enough to meet the needs of the problem.

The analysis of Table 1 and the analysis of Table 2 each satisfy Condition 1; that is for each analysis the cases are EEE. Whether Condition 2 is satisfied or not depends upon the problem at hand. We shall see how many such problems are handled as the text progresses.

One thing is clear. If a problem makes sense for a rough analysis of the cases, it will still make sense for any finer analysis. Nevertheless, the converse is not always true. That is, some problems which make sense for a fine analysis may not make sense for a rougher analysis.

4.6 Conclusions

Blaise Pascal wrote: Nature acts by progress, *itus et reditus*. It goes and returns, then advances further, then twice as much backwards, then more forward than ever, etc.

We have seen that probability theory is concerned with the building of mathematical models of unsure situations. These models are called probability models.

An unsure situation is identified by

(1) its circumstances

(2) its possible results, or cases.

Whenever the circumstances of an unsure situation are fulfilled, one and only one case happens. Thus the cases are EEE:

E for elemental: part of a case is not admitted.

E for exhaustive: one case happens.

E for exclusive: only one case happens.

In the study of an unsure situation we must first make an analysis of the cases. Our analysis must satisfy

Condition I: The cases are EEE

Condition 2: The analysis is fine enough to meet the needs of the problem at hand.

We must recognize that there is always a certain amount of inherent arbitrariness in any analysis of cases. This stems from the fact that generally we can never evaluate or imagine all the things that might happen in a given situation. As a practical matter we must be content to do the best we can for any scientific situation.

Pascal wrote: The greatness of man is great in that he knows himself to be miserable. A tree does not know itself to be miserable. It is then being miserable to know oneself to be miserable; but it is also being great to know that one is miserable.

CHAPTER 5. MATHEMATICAL FORMULATION

Blaise Pascal wrote: Clarity of mind means clarity of passion, too; this is why a great and clear mind loves ardently and sees distinctly what it loves.

5.1 Mathematical science vs. empirical science.

The state of affairs in probability theory is analogous to that in other scientific disciplines, for example geometry. As geometry is employed in everyday life, the terms point, straight line, plane, angle, etc. designate certain physical configurations. Consequently the propositions of geometry, such as the sum of the angles of a triangle is equal to 180°, formulate relations between physical configurations that can be measured.

Contrasted to the geometry used in everyday life, there is the geometry as put forth by Euclid. This geometry which consists of axioms and theorems is the geometry studied in high school mathematics. In Euclidean geometry, the terms point, straight line, plane, angle, etc. appear as undefined things, and the geometry is concerned solely with relationships among them. The axioms state certain rules which these undefined things must obey, such as two points determine a straight line. From these axioms various theorems are derived, such as the sum of the angles of a triangle is equal to two right angles. But the derivations of geometric theorems from the axioms does not depend upon the likeness that might be established between on the one hand the undefined terms like point and straight line which appear in the axioms and on the other hand the terms like point and straight line which appear in everyday life as material physical configurations.

Thus Euclidean geometry is not a branch of physics but instead is a part of conceptual mathematics. The undefined terms in Euclidean geometry such as point and straight line are not interpreted as physical entities within its framework. However, it is possible to introduce so-called coordinating definitions which make each undefined term, such as straight line in Euclidean geometry, correspond to a term employed to designate empirical subject matter, such as straight line in everyday geometry.

This distinction between Euclidean geometry and the geometry of everyday life, and the possibility of coordinating the two, is characteristic of scientific method.

On one hand, Euclidean geometry is a mathematical science. It has certain axioms, and the theorems are derived from these axioms by logical deductions. The axioms are nothing more than a set of rules about undefined things, such as point and straight line. It is meaningless to talk about the definition or true nature of such undefined things, as the mathematical structure does not care about what a point and a straight line really are.

On the other hand, the geometry of everyday life is an empirical science. It has certain findings that have been arrived at as the result of physical observation and measurement. The terms that are employed, such as point and straight line, represent material physical configurations.

There are advantages to distinguishing between mathematical structure and empirical science. One advantage is that we separate questions about mathematical validity from questions of empirical fact. Another advantage is that we increase the applicability of mathematics.

Thus, on one hand, the same formal mathematical structure may be applied to one empirical science by means of one set of coordinating definitions, and to another empirical science by means of another set of coordinating definitions. For example, the mathematical structure of the algebra of sets (or Boolean algebra) may be coordinated with many empirical sciences. We shall have more to say about Boolean algebra as the text progresses.

On the other hand, different formal mathematical structures may be applied to the same empirical science. For example, Euclidean geometry as well as several different non-Euclidean geometries may be applied to astronomy. One benefit is that it often turns out that one mathematical structure is a more effective means than another for organizing the materials of an empirical science.

In fact, there are always many benefits in the applications of mathematics to empirical sciences. There is a constant interplay between mathematical theory and empirical applications. Empirical results open up new areas for mathematical research, whereas mathematical advances open new fields for empirical investigation. Thus there is a synthesis of mathematical and practical thought. The mathematician must learn about the physical essence of a scientific problem and then find some suitable mathematical model in which to express it. The scientist must learn about the mathematical model and then look for new empirical evidence which will lead to further improvements. The end result is that both mathematics and scientific practice gain.

5.2 The applications of probability theory

Probability theory, like other branches of mathematics, has evolved out of the needs of practical application. Today these practical applications range over a tremendous number of areas in all branches of science. The tie that exists between probability theory and practical requirements has been the basic cause of the vigorous development of probability theory in recent years.

Because probability theory is applied in so many diverse fields of science, the theory must be general enough so as to provide appropriate tools for the great variety of needs. Therefore modern probability theory is not connected to any particular fields of interest, but instead represents a mathematical structure in the same way as for example Euclidean geometry does. Thus the terms used in modern probability theory, like the terms point, straight line, etc. used in Euclidean geometry, are undefined.

The central term is of course the word probability. For its mathematical usage, we must divest it of all its many connotations in the English language, and treat it instead as an undefined primitive concept.

It is unfortunate that language has no one single word for probability in its mathematical sense, but only the popular terms like chance, probability, and likelihood with their manyfold conscious and unconscious connotations. Nevertheless, the term mathematical probability or the term probability measure can be used for probability in its mathematical sense. Nevertheless these two terms are long, so we will consistently use the word probability by itself for probability in its mathematical sense.

Because probability is an undefined primitive concept, modern probability theory does not attempt to explain the so-called true meaning of probability even as Euclidean geometry does not discuss the true meaning of a straight line. Instead modern probability theory starts with a set of axioms which specify the relationship among the undefined primitive concepts. From these axioms various theorems are proved, and these theorems constitute the mathematical structure of probability theory.

This mathematical structure is applied in the form of models to practical situations. These abstract mathematical models serve as tools. Of course the same model can be used to describe different empirical situations, and different models can be used to describe the same empirical situation. The interplay between mathematical theory and empirical practice is always present. The manner in which mathematical theories are applied to practical situations depends upon experience. As our experience increases old methods of application are extended and new methods are found. The result is that both theory and practice are the benefactors.

The mathematical formulation of modern probability theory is of special importance because of the conflicting interpretations which have been given to the term probability 9 as well as the wide range of opinions concerning the conditions under which probability statements are to be regarded as significant. In the progress of the text, we will look at some of these different interpretations and opinions. Nevertheless our main objective in this book is threefold, namely

(1) to explain the axioms

(2) to derive some theorems

(3) to give many applications

.

CHAPTER 6. EXPERIMENTS

Blaise Pascal wrote: Since we cannot be universal and know all that is to be known of everything, we ought to know a little about everything. For it is far better to know something about everything than to know all about one thing. This universality is the best. If we can have both, still better; but if we must choose, we ought to choose the former. And the world feels this and does so; for the world is often a good judge.

6.1 Classification of situations as unsure

In subjects that are more or less directly concerned with specific types of objects, a clear indication of the subject matter presents no serious difficulty at the outset of a book, A reference to some of the objects can conveys certain amount of preliminary information and thereby will give the student a somewhat accurate idea of what is to follow. Such is the case in most natural sciences.

In contrast the state of affairs in probability theory is not so easy. Probability theory is not directly concerned with definite objects, but with a certain underlying characteristic common to some real-life (or real-world) situations. Situations with this characteristic are found in all branches of science. Because often such situations have little else in common, a student unacquainted with probability theory must rearrange in his mind the usual scientific classification of these situations, so as to be able to look at those possessing this unifying characteristic. This unifying characteristic is uncertainty.

For this purpose, we have classified all situations into two categories, namely sure situations and unsure situations.

We recall that a stochastic phenomenon is identified by its circumstances and its cases. In our usage the word "case" is always equivalent to the term "possible result." The cases represent the situations that can happen in an unsure situation. We recall that the cases are required to be EEE; that is

(1) E for Elemental

(2) E for Exhaustive

(3) E for Exclusive

An unsure situation has this characteristic: For each instance that the circumstances are fulfilled, one and only one case happens. In order to obtain a clear understanding of probability theory, one must first have a clear grasp of what constitute unsure situations.

6.2 The universe of an unsure situation

In this chapter we want to give further examples of unsure situations, so as to get a firm grasp on this concept.

One identifying feature of an unsure situation is that it has many possible results, or cases. Since we will be speaking about the cases of an unsure situation many times, let us give a name to the collection of all the cases. Since these cases are elementa , exclusive, and exhaustive, it is appropriate to call the set of all cases the "universal set," or briefly "the universe," of the unsure situation.

Using this concept of universe, our examples of unsure situations in Section 3.3 may alternatively be described as follows:

1. Circumstances: A coin is, tossed. Universe: heads, tails.

2. Circumstances: A die is tossed. Universe: 1, 2, 3, 4, 5, 6.

3. Circumstances: A card is drawn from the top of a pack of 52 cards. Universe: Ace of spades, Ace of clubs, etc., until all the 52 cards are listed.

4. Circumstances: A rain drop falls on a piece of paper. Universe: The points on the paper.

5. Circumstances: A seed falls from a tree and lands in a meadow. Universe: The points in the meadow.

6. Circumstances: An arrow is shot at a target. Universe: The points on the target.

6.3 Further-examples of unsure situations

Pascal wrote: Imagination disposes of everything; it creates beauty, justice, and happiness, which are everything in this world.

The concept of "an unsure situation" is a fundamental concept which we must understand as clearly as possible. The actual result of an unsure situation is a happening (or an occurrence). It is not an occurrence that can be exactly fixed in advance or perfectly predicted, but instead it is an occurrence that depends in some way and to some extent upon chance. All that we mean is that the actual result of an unsure situation is uncertain. This uncertainty as a rule is not entirely unlimited, but only prevails in certain directions and up to certain points.

For example, the statement "A man attains some age before death" represents an unsure situation. The age to which a man lives cannot be exactly fixed in advance. No person can calculate what may be the length of any particular life. The hour of a man's death is not predetermined within man's knowledge. When death comes, it happens. This happening is the result of the unsure situation. The universe that represents his age consists of all the numbers from 0 as the lower limit to say 120 years as the upper limit, as we can feel perfectly certain that his life will not stretch out more than 120 years. Thus the uncertainty about a man's final age is not entirely unlimited, but can be said to lie within a universe extending from 0 to 120 years.

We toss a penny into the air. It happens to fall heads or it happens to fall tails. This tossing represents an unsure situation. Its universe is the set "heads, tails," so that the uncertainty here is limited to two cases.

Suppose that we measure the height to which a (normal) man grows. His height will of course lie between certain extremes which delineate the universe, say between 1 meter and 3 meters. The height to which he grows represents an unsure situation. His ultimate height is a happening; that is, an occurrence dependent to some extent upon unknown and unobtainable factors.

As another example, consider the sex of a newborn baby. The baby may happen to be a boy or happen to be a girl. Its sex represents an unsure situation. The universe consists of two cases; namely, "male female."

One point should be made very clear. Some of the previous examples should have brought this point out, but anyhow it is worth stressing again. The point is this: The role played by chance in an unsure situation is not required to be complete but may be to any degree.

For example, when a coin or a die is tossed, we usually regard the role played by chance as being entire, and the role played by the player as having no effect on the ultimate result. Nevertheless, there may be individuals with extraordinary skill who can control some of the conditions in tossing the coin, so that they have some influence on the way it falls.

On the other hand, consider the act of shooting an arrow at a target. Because the man takes aim, the ultimate result to some extent depends upon his skill. Nevertheless, this act is an unsure situation because the exact point where the arrow strikes is uncertain. The role played by the man has some effect on the ultimate result, and the role played by chance has some effect. Thus chance may be regarded only as a co-agent in the final result. In fact, most stochastic phenomena are of the type in which chance may be regarded only as a co-agent in the ultimate result.

6.4 Experiments

We have classified "situation" into two categories; namely

(1) Sure situation

(2) Unsure situation

There are other classifications. In particular, we may classify "situation" into the following two categories.

(1) Experimental situation

(2) Non-experimental situation

An experimental situation is one which can be repeated for an indefinite number of instances, each under identical circumstances. In each instance, the same conditions are fulfilled and one result is produced. Thus if the experimental situation is repeated five times, a series of five results is produced. All other situations fall into the categories of non-experimental situations.

Situations that are both unsure and experimental are very useful in gaining insight to the methods of probability theory. For brevity, we shall replace the long expression "experimental unsure situation" by the single word "experiment." Thus an experiment is an unsure situation that can be repeated indefinitely, where each repetition is made under the same circumstances and produces one result. Twenty repetitions of the experiment would produce a series of twenty results. Each result in the series, of course, must be one and only one of the possible results, or cases, of the unsure situation.

An example of an experiment is the tossing of a coin. This experiment can be repeated an indefinite number of times. For each repetition, either H or T happens. Thus a series produced by 20 repetitions might be

HHHTTHHTHTHHTTTUTHTT

Another example would be the game of roulette. The same conditions hold for each spin of the wheel. Each legitimate game of chance, such as dice or cards, represents a means of doing an experiment. The intuitive concept of experiments repeated under identical conditions leads to the notion of stochastic independence. When a scientist says that two experiments are performed under identical conditions, he implies independence. In other words, the scientist infers that the result of either experiment has no influence on the result of the other experiment. For example, if we toss a coin twice under identical conditions, then the two tosses are independent. The appearance of heads or tails on toss 1 does not influence which side appears on toss 2

The notion of stochastic independence applies not only to repetitions of the same experiment under identical conditions, such as the successive tossing of a coin, but also applies to different experiments that have nothing to do with each other. Such experiments can either be done simultaneously (i.e. at the same time) or sequentially (i.e. in a time succession).

For example, suppose we toss a coin and throw a die, either simultaneously or sequentially. It seems evident that the fall of the coin has nothing to do with the fall of the die. Accordingly, we say that these two experiments are independent

CHAPTER 7. NOTION OF PROBABILITY

Blaise Pascal wrote: Poetical beauty. ...We know well what is the object of mathematics, and that it consists of proofs, and what is the object of medicine, and that it consists of healing. But we do not know in what grace consists, which is the object of poetry.

7.1 The ratio of the number of boys to the total number of births

An unsure situation well known to everyone is "A baby is to be born. What is its sex? Science up to this point has not found a way to predict the sex of a newborn baby at the time of conception. The biological mechanism that determines the sex is not understood, and research shows that this problem is exceedingly complex. However, the possible results of the phenomenon are very simple, namely the result must be one of two alternatives; namely boy or girl. (Note. We treat the birth of twins as 2 separate births, the birth of triplets as 3 separate births, etc.)

Suppose that all newborn babies are registered in order as they are born in a country such as the United States for a given year. Boys and girls (B and G) follow each other without apparent regularity, as

BBGGGBGGGBGGBGB

Each birth represents the outcome of an unsure situation, and the series of all births represent repeated trials of this basic unsure situation. We cannot predict the result of any one given birth, whether it will be a boy or a girl. Accordingly, we cannot predict the details of this series. Nevertheless we can predict an important overall feature of the series by summing the number of boys born and by summing the number of girls born. The number of boys turns out to be greater than the number of girls. In fact, these two numbers will be very closely proportional to 51.5 : 48.5.

The series of registrations of births represents statistical data. The knowledge that approximately 51.5 percent of the babies are boys, and 48.5 percent of the babies are girls, represents empirical experience drawn from the statistical data. Thus in ancient times it was observed

that the ratio of male births to the total number of births in entire
countries remained almost unchanged from year to year. This ratio was
found to be approximately 1/2 by the Chinese over 2200 years ago. How
did the ancient Chinese attain the empirical knowledge that the sex
ratio is approximately 50 percent? They arrived at it by means of
statistical data collected by censuses of their population.

In modern times, of course, statistics are kept much more accurately,
and so today we know the sex ratio more precisely. Thus out of a large
number of births, approximately 51.5 percent will be boys and 48.5
percent will be girls. It is important to bear in mind that over the years
the sex ratio has been determined from the available statistical data of
actual births, and not by theoretical reasoning. Thus out of many births,
the number of boys and the number of girls are counted. Such numbers
form the pertinent statistical data.

For example, the statistics for the United States in the year 1943 show
that there were 1,506,959 boys born and 1,427,901 girls born out of a
total of 2,934,860 births. We recognize the numbers

1,506,959	frequency of male births
1,427,901	frequency of female births
2,934,860	Total births

From these data we compute the ratios

relative frequency of male births	$\dfrac{1,506,959}{2,934,860} = 0.513$
relative frequency of female births	$\dfrac{1,427,901}{2,934,860} = 0.487$

As another example, the statistics for the United States in the year 1950
show that there were 1,823,555 boys born and 1,730,594 girls born, out
of a total of 3,554,149 births. From this data, the ratio of the number of
boys to the total number of births is the relative frequency

$1,823,555/3,554,149 = 0.513$

This relative frequency says that 51.3 percent of the babies born in the United States in 1950 were boys.

7.2 The stability of the relative frequency

Our empirical experience is that 51.5 percent of births are boys. A remarkable feature of this relative frequency is its stability; that is, it does not noticeably change from place to place and from year to year. Of course there is some fluctuation, but this fluctuation is relatively small.

The lesson we draw is this. Statistical data such as the number of boys and girls born in the United States in various years establishes and supports our empirical knowledge about relative frequencies. Thus although the sex of any particular birth is unknown in advance and so represents an unsure situation, our accumulated empirical experience tells us that out of 100 births approximately 51.5 percent will be boys and 48.5 percent will be girls.

From statistical evidence, the empirical experience of mankind grows, thereby making the study of unsure situations more amenable and the uncertainties of anticipation more bearable. For each individual birth we are unable to predict whether the baby will be a boy or a girl. But the accumulated empirical experience of mankind tells us that, out of a large number of new born babies, approximately 50 percent will be boys, so that the chance of a boy is approximately 50 percent, regardless of the country or the year.

The stability of the statistical ratio of sex is well understood and appreciated by mankind. We might say that our entire social structure is based on this stability. Think of the repercussions that would result if for some generation or for some countries they were approximately twice as many girls born as boys, or vice versa.

7.3 The probability of a boy

Despite the fact that there are many babies being born all the time, we cannot get away from the fact that any particular birth is an unsure situation by itself. Its stochastic nature does not depend upon the fact that other births are taking place. We can always look at any particular birth by itself, isolated from other births. The fact remains that the sex of a particular birth is unsure. The birth of either a boy or a girl may take place.

Let us consider an event associated with an unsure situation. The fundamental notion is that this event has a probability. This probability is a number, between 0 and 1; that is, a measure of the chance of an event happening. This number is called "the probability that the event happens." When the context is clear, this number is called "the probability of the event" or simply "the probability."

More specifically, consider the event "a boy" for the stochastic phenomenon "birth". This event has a probability, which we call "the probability that the event of being a boy happens" or simply "the probability of a boy."

To illustrate the usage of mathematical notation, we may let the letter B designate the event of "a boy." Then we may let the symbol P(B) (which is read "P of B") designate the probability of a boy. The probability P(B) is a number. This number must be a number on the interval between 0 and 1 including the end points 0 and 1. In other words, the probability must not be less than 0 nor greater than 1.

But what is the value of this number? No one knows exactly. But the notion of this probability is natural to each of us, and through experience and reflection we may obtain what we consider a reasonable evaluation of its numerical value.

As we have seen, mankind has not yet been successful in the use of theoretical methods to evaluate this probability. Nevertheless through experience, we do know that the 51.5 percent of all births are boys. In other words, we know that the ratio is 0.515. On this basis we have reason to place the probability of a boy to be 0.515. In mathematical notation, we have

$P(B) = 0.515$

There are two means of evaluating probabilities of events:

(1) By accumulated experience

(2) By theoretical reflection and argument.

In the case of the probability of a boy, means (1) was used. Nevertheless a breakthrough in biological research might allow us also to use means (2).

> Pascal wrote: When it is said that heat is only the motions of certain molecules, and light the *conatus recedendi* which we feel, it astonishes us. What! Is pleasure only the ballet of our spirits? We have conceived so different an idea of it! And these sensations seem so removed from those others which we say are the same as those with which we compare them! The sensation from the fire, that warmth which affects us in a manner wholly different from touch, the reception of sound and light, all this appears to us mysterious, and yet it is material like the blow of a stone. It is true that the smallness of the spirits which enter into the pores touches other nerves, but there are always some nerves touched.

7.4 The probability of a face on a 20-sided die

In Euclidean geometry, a Platonic solid is a regular, convex polyhedron. The Platonic solids have an aesthetic beauty. The faces on each of them are congruent, regular polygons, with the same number of faces meeting at each vertex. There are exactly five solids which meet these criteria; each is named according to its number of faces. The tetrahedron has four faces. The cube or hexahedron has six faces. A die is a cube with the faces numbered. The octahedron has eight faces. The dodecahedron has twelve faces. The icosahedron has twenty faces.

Let us now illustrate the use of means (2), namely, the use of theoretical reflection and argument, to evaluate a probability. Let the stochastic phenomenon be the tossing of a homogeneous and symmetrical 20-sided die. In other words, the die used here is an icosahedron with its sides numbered from 1 to 20. There is no accumulated empirical experience on this phenomenon. Nevertheless we do have much accumulated experience on the tossing of a 6-sided die which very accurately meets the ideal homogeneous and symmetrical properties. Our experience there is that the probability of each face is 1/6. By theoretical reflection and argument we may therefore evaluate the probability of each face of a homogeneous, symmetrical 20-sided die to be 1/20.

7.5 The verity of the science of probability

The modern development of the science of probability is associated with the increased interest in the mathematical theory as well as a wide broadening of its applications to practical situations. The modern development of the mathematical theory of probability was initiated by George Boole who connected the theory of probability with symbolic reasoning. Boole's classic work "An Investigation of the Laws of Thought, on which are founded the Mathematical Theories of Logic and Probabilities," published in 1854, marks the beginning of modern probability theory. Boole's work has led to the axiomatic formulation of probability theory. Today the axiomatic method is recognized as the clearest approach to any mathematic discipline, and it is indispensable for the understanding of the nature of probability and the state of its foundations. The axiomatic concept of the science of probability has permitted the elucidation of many philosophical difficulties as well as the advancement of the science itself.

Together with the mathematical development, there has been a correspondingly vigorous development of the applications of probability theory to practice. We must always keep in mind that it is its concrete contact with reality and its practical utility that make the science of probability what it is.

The verity of science of probability is twofold. There is the verity that results from knowing that all the theorems are deduced logically from the definitions and axioms. There is also the verity that results from concrete experience

> Blaise Pascal wrote: When we would pursue virtues to their extremes on either side, vices present themselves, which insinuate themselves insensibly there, in their insensible journey towards the infinitely little; and vices present themselves in a crowd towards the infinitely great, so that we lose ourselves in them, and no longer see virtues. We find fault with perfection itself.

CHAPTER 8. HEADS AND TAILS

Blaise Pascal wrote: A game is on, and heads or tails will turn up. What will you wager? According to reason you cannot leave either; according to reason you cannot leave either undone... Yes, but wager you must; there is no option, you have embarked on it. So which will you have? Come. Since you must choose, let us see what concerns you least. You have two things to lose: truth and good, and two things to stake: your reason and your will, your knowledge and your happiness. And your nature has two things to shun: error and misery. Your reason does not suffer by your choosing one more than the other, for you must choose. That is one point cleared. But your happiness? Let us weigh gain and loss in calling heads that God is. Reckon these two chances: if you win, you win all; if you lose, you lose naught. Then do not hesitate, wager that He is.

8.1 Tosses of a coin

A coin has 2 sides: one called heads which is abbreviated by H, and the other side is called tails which is abbreviated by T.

(0 tosses). A coin is not tossed into the air. In tossing the coin no times, nothing can happen. The set that has nothing is the empty set { }. The event in question is the empty set which contains nothing at all. We can represent this result by the rectangular block:

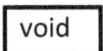

(1 toss). A coin is tossed once into the air and the side facing up is recorded. There are two possible results, namely H and T. This set {H, T} of the 2 possible results is EEE; that is, elemental, exhaustive, and exclusive. The events in question are two sets; namely, 1H, 0H. We can represent this result by two blocks, which we place below the block we already have. We obtain

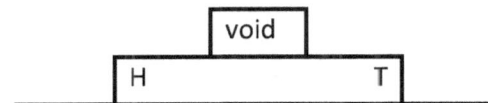

(2 tosses). A coin is tossed twice into the air and each time the side facing up is recorded. We see that the 2 tosses can lead to 4 possible results; that is, HH, HT, TH, TT. Each of these 4 possible results is an

event. Thus the possible result HT represents the event "toss 1 gives H and toss 2 gives T." Any event that consists of just 1 possible result is called a "simple event." Any event that consists of more than one possible result is called a "compound event." Hence the event consisting of the possible result HT is a simple event. Instead of the 4 simple events above, we may also consider other events. As we have seen the simple event "toss 1 is H and toss 2 is T" consists of the possible result HT. Similarly the simple event "toss 1 is T and toss 2 is H" consists of the possible result TH. But the event "one H and one T without regard to order", or in other words "1 of the 2 tosses is H", consists of both of the possible results HT and TH. That is, this event is the set {HT, TH} and so is a compound event. The events in question are three sets; namely 2H, 1H, 0H. We can represent this result by three blocks, which we place below the blocks we already have. We obtain

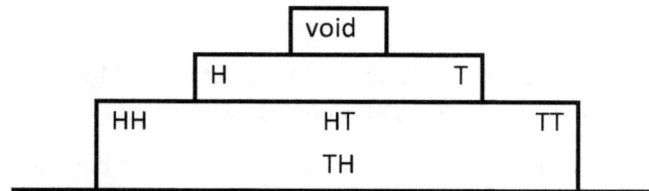

(3 tosses). A coin is tossed thrice into the air and each time the side facing up is recorded. There are 8 possible results; namely, HHH, HHT, HTH, THH, HTT, THT, TTH, TTT. Each of these 8 possible results is a simple event. The events in question are four sets; namely, 3H, 2H, 1H, 0H.We can represent this result by four blocks, which we place below the blocks we already have. We obtain

(4 tosses). A coin is tossed four times into the air and each time the side facing up is recorded. The 16 possible results are: HHHH, HHHT, HHTH, HTHH, THHH, HHTT, HTHT, HTTH, THHT, THTH, TTHH, HTTT, THTT, TTHT, TTTH, TTTT. The events in question are five sets; namely 4H, 3H, 2H, 1H, 0H.

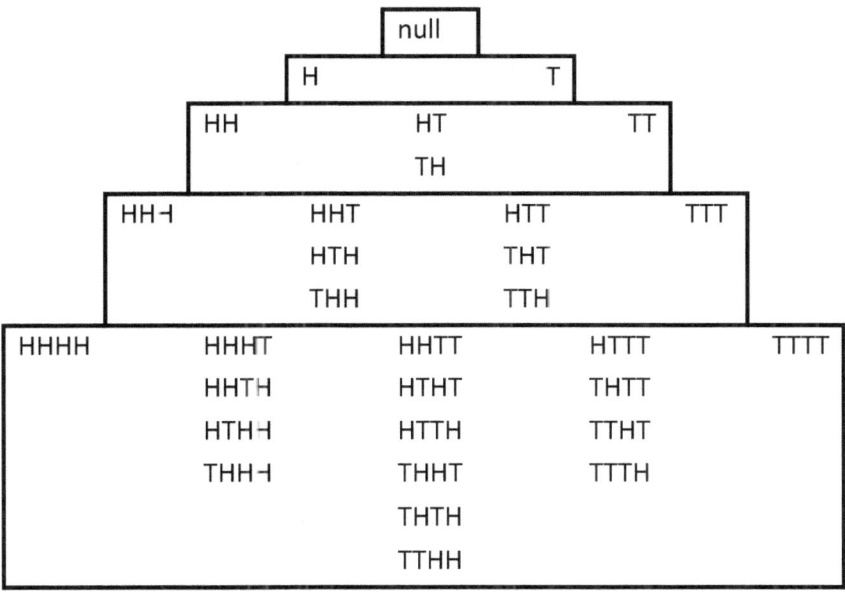

Count the number of number of entries in each block, and replace the entries by the number counted. We obtain

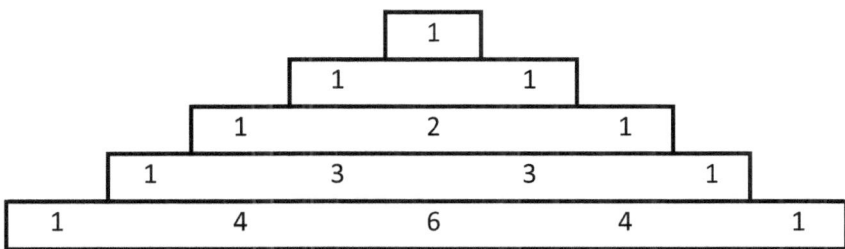

This triangular array is called Pascal's arithmetic triangle. More and more rows can be added to Pascal's triangle.

Look at the triangle. Can you discover the rule for finding the entries in the arithmetic triangle? The rows are numbered starting at 0 (for zero tosses). It has entry 1. The rule is this: In the 1 (for 1 toss), write the number 1 repeated 2 times. There are no other entries in this row. Then the entry in any other row is found by adding together two entries in the above row; namely, the entry just to the left and the entry just to the right.

Let us do row 2, which has the entries 1, 2, 1. To find the first entry in row 2, we note that there no entry above just to the left. The entry above just to the right is 1. The sum is 0 + 1 = 1, which is the first 1 in row 2. The 2 in row 2 is found by adding the two above entries; that is, is 1 + 1 =2. The second 1 in row 2 is found by adding the two above entries; that is, is 1 + 0 =1.

Let us do row 3, which has the entries 1, 3, 3, 1. We have:

    0 + 1 = 1
    1 + 2 = 3
    2 + 1 = 3
    1 + 0 = 1

Let us do row 4, which has the entries 1, 4, 6, 4, 1. We have:

    0 + 1 = 1
    1 + 3 = 4
    3 + 3 = 6
    3 + 1 = 4
    1 + 0 = 1

Let us do row 5, whose entries we do not yet know. We have:

    0 + 1 = 1
    1 + 4 = 5
    4 + 6 = 10
    6 + 4 = 10
    4 + 1 = 5
    1 + 0 = 1

Let us do row 6, whose entries we do not yet know. From above, we see that 1, 5, 10, 10, 5, 1 are the entries of row 5. By adding adjacent entries, we obtain of we find the entries of row 6, which are 1, 6, 15, 20, 15, 6, 1.

By use of our rule we may prolong the arithmetic triangle to any number of tosses that we wish.

The generating function provides another way to find the entries in Pascal's triangle. The generating function for the triangle is $1 + x$. We create the table:

Row	Polynomial	Expansion of polynomial
0	$(1+x)^0$	1
1	$(1+x)^1$	$1 + x$
2	$(1+x)^2$	$1 + 2x + x^2$
3	$(1+x)^3$	$1 + 3x + 3x^2 + x^3$
4	$(1+x)^4$	$1 + 4x + 6x^2 + 4x^3 + x^4$
5	$(1+x)^5$	$1 + 5x + 10x^2 + 10x^3 + 5x^4 + x^5$

We see that the coefficients of the expanded polynomials give Pascal's triangle. If we let x = 1, we obtain

Row	Polynomial
0	$(1 + 1)^0 = (2)^0 = 1$
1	$(1 + 1)^1 = (2)^1 = 2$
2	$(1 + 1)^2 = (2)^2 = 4$
3	$(1 + 1)^3 = (2)^3 = 8$
4	$(1 + 1)^4 = (2)^4 = 16$
5	$(1 + 1)^5 = (2)^5 = 32$

The above table shows that the total of the entries in the row n is 2^n. For example the total of the entries in row 3 s

$$1 + 3 + 3 + 1 = 2^3 = 8$$

The total of the entries in row 4 is

$$1 + 4 + 6 + 4 + 1 = 2^4 = 16$$

In other words, the total doubles each time that we go from one row to the next row. We expect this because each new toss of the coin doubles the number of possible results. Also we should remember that the number of non-zero entries in the row n is equal to n+1. Thus for n = 4 tosses there are 5 entries; namely, 1, 4, 6, 4, 1.

In addition we should remember that the columns in the arithmetic triangle are symmetrical, that is, the entries equally distant from the extremes are equal. In the rows for which the number n of tosses is an even number, the magnitude of the entries increases up to the middle entry. This greatest entry is for the event in which the number of H's and the number T's are equal. For example, for n = 4, we have the entries

1, 4, 6, 4, 1

The greatest entry 6 is for Event 2H, which is made up of 2 H's and 2 T's.

In the rows for which the number n of tosses is an odd number, the equal division of H's and T's cannot be realized. Thus in the middle of the row, there are 2 equal entries, which are greater than the other entries. These greatest entries correspond to the two events where one of the 2 sides of the coin comes up one time more than the other. For example, for n = 5 we have the entries

1, 5, 10, 10, 5, 1

One of the greatest entries 10 is for Event 2 H, that is, 2 H's and 3 T's. The other is for Event 3 H, that is, 3 H's and 2 T's.

8.2 Probabilities involved

(0 tosses). The coin is not tossed. The empty event { } always occurs when you do not toss the coin. As a result the empty event has a probability of one; that is, for no tosses of a coin, P({ }) = 1. We have:

No toss	possible result	probability
void	void	P(void) = 1

The following shows the structure of no event:

Event	Description	Set	Probability
nothing	No toss	{ }	P({ }) = 1

(1 toss). In tossing the coin one time, there are two simple events; namely, H or T. Each of the two events H or T has a probability. We shall make the hypothesis that the coin is fair. In such a case, H has the chance of 1 in 2, so that the probability of H is 0.5. Likewise, T has 1 chance in 2, so that the probability of T is 0.5. This equally-likely hypothesis, of course, represents an ideal. However, it can be sufficiently realized in practice so that the conclusions to which it leads are in accord with experience. We have

toss 1	possible result	probability
H	H	P(H) = 1/2
T	T	P(T) = 1/2

The following shows the structure of two events:

Event	Description	Set	Probability
1H	1 of the 1 toss is H	{H}	P(1H) = 1/2
0H	0 of the 1 toss is H	{T}	P(0H) = 1/2

(2 tosses). In tossing the coin twice, there are four simple events. Toss 1 can result in H or T. Toss 2 can result in H or T. Toss 2 is in no way influenced by the outcome of toss 1. In other words, the result of toss 1 has no influence on the result of toss 2. We may say that the coin has neither mind nor memory. The outcome of toss 2 is independent of the outcome of toss 1. This point cannot be emphasized too much. One must fully understand it in order to appreciate what follows. Each of the four simple events has 1 chance in 4 to happen. Hence its probability is 1/4.

By combining the possible results of toss 1 with the possible results of toss 2, we get:

toss 1	toss 2	possible result	probability
H	H	HH	P(HH) = 1/4
H	T	HT	P(HT) = 1/4
T	H	TH	P(TH) = 1/4
T	T	TT	P(TT) = 1/4

The following table shows the structure of three events:

Event	Description	Set	Probability
2H	2 of the 2 tosses are H	{HH}	P(2H) = 1/4
1H	1 of the 2 tosses is H	{HT, T H}	P(1H) = 2/4
0H	0 of the 2 tosses is H	{TT}	P(0H) = 1/4

Event 2H is a simple event. Event 1H is a compound event. Event 0H is a simple event. The probabilities of these 3 events are not equal, because Event H, namely "1 of the 2 tosses is H," consists of 2 different possible results, namely HT and TH. Hence the chances of this event are 2 in 4, so that its probability is 2/4. On the other hand, Event 2H consists of 1 possible result, namely HH. Hence the chance of this event is 1 in 4, so that its probability is 1/4. Similarly, Event 0H is consists of 1 possible result, namely TT. Hence its chance is 1 in 4, so that its probability is 1/4.

(3 tosses). In three tosses of a coin, there are 8 simple events. The chance of each of the simple events is 1 in 8, so the probability of each is 1/8.

toss 1	toss 2	toss 3	possible result	probability
H	H	H	HHH	1/8
H	H	T	HHT	1/8
H	T	H	HTH	1/8
H	T	T	THH	1/8
T	H	H	TTH	1/8
T	H	T	THT	1/8
T	T	H	HTT	1/8
T	T	T	TTT	1/8

Event 3H has 1 chance in 8 to happen. Event 2H has 3 chances in 8 to happen. Event 1H has 3 chances in 8 to happen. Event 0H has 1 chance in 8 to happen. In summary, the 4 events have the probabilities:

Event	Description	Set	Probability
3H	3 of the 3 tosses are H	{HHH}	P(3H) = 1/8
2H	2 of the 3 tosses are H	{HHT, HTH, THH}	P(2H) = 3/8
1H	1 of the 3 tosses is H	{HTT, HHT, TTH}	P(1H) = 3/8
0H	0 of the 3 tosses is H	{TTT}	P(0H) = 1/8

(4 tosses). In four tosses of a coin, there are 16 simple events. The chance of each of the simple events is 1 in 16, so the probability of each is 1/16. The 5 events 4H, 3H, 2H, 1H, 0H have the probabilities:

Event	Description	Set	Probability
4H	4 of the 4 tosses are H	{HHHH}	P (4H) = 1/16
3H	3 of the 4 tosses are H	{ HHHT, HHTH, HTHH, THHH }	P (3H) = 4/16
2H	2 of the 4 tosses are H	{HHTT, HTHT, HTTH, THHT, THTH, TTHH}	P (2H) = 6/16
1H	1 of the 4 tosses is H	{HTTT, THTT, TTHT, TTTH}	P (1H) = 4/16
0H	0 of the 4 tosses is H	{TTTT}	P (0H) = 1/16

In the same way we may continue to any number of tosses, and find the probabilities of the various events.

8.3 Fibonacci sequence

As we have seen, in order to construct Pascal's Triangle, we started with an apex of 1. Every number below in the triangle is the sum of the two numbers diagonally above it to the left and the right, with positions outside the triangle counting as zero. Pascal's triangle is

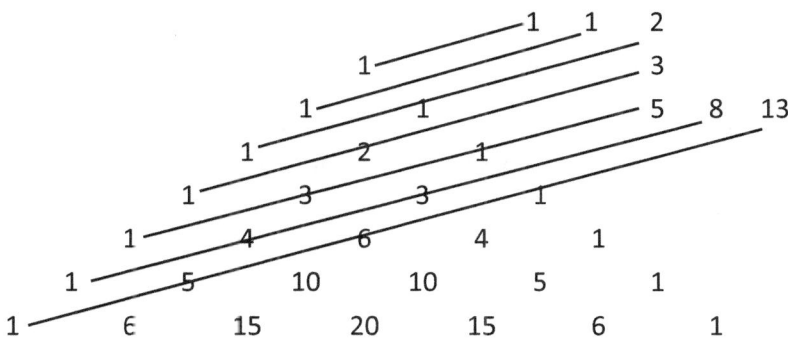

The slanting lines represent the "shallow diagonals". The numbers on them add to the Fibonacci series 1, 1, 2, 3, 5, 8, 13, ...

1=1	1=1	1+1=2	1+2=3
1+3+1=5	1+4+3=8		1+5+6+1=13

8.4 Some problems and solutions

Problem 1. Which is more probable: Event A = {3 H's in 4 tosses} or Event B = {5 H's in 8 tosses}?

Solution. Event A is Event 3H for 4 tosses. Hence from the arithmetic triangle, we see that event A has 4 chances in 16, so

P(A) = 4/16 = 8/32

We also see that event B is Event 5 H for 8 tosses. Hence event B has 56 chances in 256, so

P(B) = 56/256 = 7/32

Thus event A is more probable than event B.

Problem 2. Which event is more probable: A = {at least 3 H's in 4 tosses} or B = {at least 5 H's in 8 tosses}?

Solution. Event A is the union of the 2 events:

Event 3 H	3 H's in 4 tosses	4 chances in 16
Event 4 H	4 H's in 4 tosses	1 chance in 16

The chances were found from the arithmetic triangle. We see that

Thus P(A) = 5/16 = 80/256.

Event B is the union of the 4 events:

Event 5H	5 H's in 8 tosses	56 chances in 256
Event 6H	6 H's in 8 tosses	28 chances in 256
Event 7H	7 H's in 8 tosses	8 chances in 256
Event 8H	8 H's in 8 tosses	1 chance in 256

The event B has 56 + 28 + 8 + 1 = 93 chances in 256 to happen, so

P(B) = 93/ 256. Since P(A) = 80/256, we see that event B is more probable than event A.

CHAPTER 9. EXPECTATION

Pascal wrote: Let man consider what he is in comparison with all existence; let him regard himself as lost in this remote corner of nature; and from the little cell in which he finds himself lodged, I mean the universe, let him estimate at their true value the earth, kingdoms, cities, and himself. What is a man in the Infinite?

But to show him another prodigy equally astonishing, let him examine the most delicate things he knows. Let a mite be given him, with its minute body and parts incomparably more minute. Dividing these last things again, let him exhaust his powers of conception, and let the last object at which he can arrive be now that of our discourse. Perhaps he will think that here is the smallest point in nature. I will let him see therein a new abyss. I will paint for him not only the visible universe, but all that he can conceive of nature's immensity in the womb of this abridged atom. Let him see therein an infinity of universes, each of which has its firmament, its planets, its earth, in the same proportion as in the visible world; in each earth animals, and in the last mites, in which he will find again all that the first had, finding still in these others the same thing without end and without cessation. Let him lose himself in wonders as amazing in their littleness as the others in their vastness. For who will not be astounded at the fact that our body, which a little while ago was imperceptible in the universe, itself imperceptible in the bosom of the whole, is now a colossus, a world, or rather a whole, in respect of the nothingness which we cannot reach? He who regards himself in this light will be afraid of himself, and observing himself sustained in the body given him by nature between those two abysses of the Infinite and Nothing, will tremble at the sight of these marvels.

9.1 Expectation

Pascal devised the mathematical concept of expectation (or expected value). The expectation is the product of the probability of winning times the possible winnings.

expectation = probability × winnings

Let us suppose that a person judges that the probability of his winning as 3 chances in 10, or 0.3. If he expects to win $100 in the event of a favorable outcome, his expectation is 0.3 × $100 = $30.

Suppose the prize in a lottery is $100. A person holds 1 ticket out of 50 issued. His probability of winning is (1/50) = 0.02. His expectation is 0.02 × $100 = $2, which is the proper valuation of his ticket. If he decides to sell his ticket before the drawing, then its fair market value is $2.

The concept of expectation can be used determining when to continue or when to stop in an uncompleted venture. For example, suppose a player is faced with a choice at some point in the games. He can bet $20, which would increase his potential winnings to $80. At this point he must estimate his probability of winning. If his probability of winning is 0.4, then his expectation is 0.4 × $80 = $32. To get this expectation, he must pay $20. In other words, an expectation worth $32 can be purchased for $20. Because the expectation is worth more than the purchase price, he should bet the $20.

On the other hand, if his probability of winning is 0.1, then his expectation is 0.1 × $80 = $8. To get this expectation, he must pay $20. In other words, an expectation worth $8 can be purchased for $20. Because the expectation is worth less than the purchase price, he should not bet the $20.

9.2 Expectation of an event

The probability of an event is not the only quantity in which we are interested in the study of a stochastic phenomenon. For example, it is very different to have 1 chance in 10 to gain $100 or to have 1 chance in 10 to gain $100,000. This consideration leads to the notion of the "mathematical expectation of an event." A shorter term is the "expectation of an event."

If a person has 1 chance in 10 to gain $100, then his expectation is

$100 (1/10) = $10

If a person has 1 chance in 10 to gain $100,000, then his expectation is

$100,000 (1/10) = $10,000

If a person has 1 chance in 2 to gain $200, then his expectation is

$200 (1/2) = $100

Suppose that the only prize given at a lottery is $100,000, and that there are 200,000 tickets sold at $1 each. Then the expectation of a ticket is

$100,000 (1/200,000) = $0.50

even though the cost of the ticket was $1.00. Examining this situation, we see that the people who organized this lottery realize $200,000 from ticket sales, and only pay out $100,000 as a prize, thereby making a profit of $100,000 On the other hand, each of the 200,000 ticket holders paid $1.00 for a ticket whose expectation was only $0.50.

The expectation of a contingent gain is the product of the gain times the probability of realizing this gain. We can say that the expectation is the value of a gain whose attainment is not certain.

When it is a matter of a loss instead of a gain, we shall consider a loss to be a negative gain For example, if a person has 1 chance in 50 to lose $100, his expectation is −$100 (1/50), which is −$2. The expectation is negative. A negative expectation is the value of a loss, whose realization is not certain but only contingent. If there is a probability of 1/20 to sustain a loss of $800, the expectation is −$800 (1/20) = −$40. In order to avoid this risk of losing $800, we may say that $40 is the "fair" amount that should be paid.

Suppose that for a toss of a coin, a player receives $100 if the coin lands H (where H stands for heads). The probability of H is 0.5. We say that the expectation of the player is $100 (0.5) =$50.

Suppose that for two consecutive tosses of a coin, another player receives $100 if both tosses result in H. The probability that both tosses land H is 0.25. The expectation of this player is then $100 (0.25) = $25.

The expectation of a player is the product of his possible gain times the probability that he has of realizing this possible gain.

Summing up, we have: The expectation of an event is the product of the gain received if the event happens times the probability that the event happens. The expectation is a fictitious or imaginary sum of money. It does not ordinarily correspond to a possible value of gain or of loss.

If a person has 1 chance in 10 to gain $40 his expectation is $4. But $4 is not a possible value of his gain. The only possible values that can happen are $0 and $40.

If a person has 1 chance in 5 to lose $10, his expectation is a loss of $2. Nevertheless, either a loss of $0 or a loss of $10 will happen, never a loss of $2.

9.3 Expectation of several events

An advantage of the notion of expectation is the following. The combination of different probabilities can be complicated, but the combination of different expectations is simple and intuitive. It is easy to see that expectations can be added together as sums of ordinary money. The rule is:

> The expectation for several events is the sum of the expectations for each of the events.

This property of addition can make the calculation of expectations very easy in many applications. Suppose-a coin is tossed 3 times. If it lands H on toss 1, a player receives $10. If it lands H on toss 2, he receives $20. If it lands H on toss 3, he receives $30. Thus we have

Event	Gain	Probability	Expectation
H on toss 1	$10	0.5	$5
H on toss 2	$20	0.5	$10
H on toss 3	$30	0.5	$15

Therefore his total expectation is

$5 + $10 + $15 = $30

A negative gain is a loss. In dealing with expectations, we must be prepared to consider negative gains (or losses) as well as positive gains.

Suppose a coin is tossed. A player receives $100 if the coin lands H and loses $100 if the coin lands T. What is his expectation? The loss of $100 is the same as a gain of −$ 100. Hence we have the table:

Event	Gain	Probability	Expectation
H	$100	0.5	$50
T	−$100	0.5	−$50

Hence his total expectation is

$50 + (−$50) = $50 −$50 = 0

9.4 Games of chance

Blaise Pascal wrote: Let man then contemplate the whole of nature in her full and grand majesty, and turn his vision from the low objects which surround him. Let him gaze on that everlasting light, set like an eternal lamp to illumine the universe; let the earth appear to him a point in comparison with the vast circle described by the sun; and let him wonder at the fact that this vast circle is itself but a very fine point in comparison with that described by the stars in their revolution round the firmament. But if our view be arrested there, let our imagination pass beyond; it will sooner exhaust the power of conception than Nature that of supplying material for conception. The whole visible world is only an imperceptible atom in the ample bosom of nature. No idea approaches it. We may enlarge our conceptions beyond all imaginable space; we only produce atoms in comparison with the reality of things. It is an infinite sphere, the center of which is everywhere, the circumference nowhere. In short it is the greatest sensible mark of the almighty power of God, that imagination loses itself in that thought.

The subject of expectation is used very much in connection with games of chance. In the game of heads or tails, each toss of the coin is an individual turn. For each turn the player calls either H or T and wins if the coin lands what he called. We assume that he coin is fair; that is, $P(H) = 0.5$ and $P(T) = 0.5$. His probability of winning is 0.5 and his probability of losing is 0.5. If we let W stand for win and L stand for lose, we may write in symbols that $P(W) = 0.5$ and $P(L) = 0.5$ for each turn of the game of heads or tails. All the developments of Chapter 8 for H and T also apply for W and L.

The game of heads or tails is a game of pure chance. There is no skill involved. Some games of chance are prohibited by law, and most are subject to certain restrictions. To apply these laws, various criteria have been put forth to distinguish games of chance from games of skill. Usually in legal works games of chance are described as "games of an aleatory character." The first definition of ALEATORY is "depending on an uncertain event or contingency as to both profit and loss. An example of usage would be "an aleatory contract." We recall that "alea" is the Latin word for die. Thus what is meant is that games of chance are games that have the characteristics of a dice game. The second definition of ALEATORY is "relating to luck and especially to bad luck."

Roulette is a game of chance whereas chess is a game of skill. The two extremes are games of pure chance and games of pure skill. Such extreme cases can be easily classified. All kinds of intermediate cases between the two extremes are possible. These intermediate kinds of games combine elements of skill together with certain elements of chance. For example, most card games fall into this intermediate category, for although they require wit, there is always at least that element of chance resulting from the shuffling of the cards before dealing.

Generally speaking, the classification games of chance includes

(1) Games of pure chance, that is, games for which the probabilities of winning for the individual players are entirely independent of their skill

(2) Games predominated by chance, that is, games for which the probabilities of winning for the individual players depend only very slightly on their skill.

From this point of view games can be divided into 3 categories:

(1) Games of pure chance (coin tossing, dice, some card games, roulette, etc.)

(2) Games of chance and skill (most card games, dominos, etc.)

(3) Games of pure skill (chess, etc.).

One cannot fully comprehend the prodigious number of different possible results that can be obtained from a pack of cards. The number of possible arrangements of 52 cards is expressed by an 8 followed by 67 other digits.

In a game of coin tossing, dice, or roulette, a player can have in his mind the probabilities of the different events. Nevertheless, it is generally impossible for a card player to have in his mind, in a complete way, all of the different probabilities. The interest in card games is not diminished. The player not being able to completely know all the probabilities must appeal to his experience and intuition to supplement his partial knowledge. Thus his skill and cleverness play a part, and the game no longer appears as completely dependent upon chance.

On the other hand, coin tossing, dice, and roulette depend completely upon chance. Since it is easy to know the probabilities in every situation, there is no opportunity for a player to make use of any skill or cleverness. A knowledgeable player is one who knows these probabilities and acts on their basis. On the other hand, an unknowledgeable player is one who does not know the probabilities and is under the illusion that his skill and cleverness play a part. (Here we rule out the use of skill and cleverness as a means to cheat.) A knowledgeable player always has an advantage over a less knowledgeable player.

9.5 Advantageous, equitable, and disadvantageous games

A game is "advantageous" to a player if his expectation is positive. It is "disadvantageous" if his expectation is negative. It is neither advantageous nor disadvantageous when his expectation is zero. We then say that the game is "equitable."

For example, suppose a player has 1 chance in 3 to gain $24 and 2 chances in 3 to lose $12. This game is equitable because the expectation of the player is

$$\$24 \ (1/3) - \$12 \ (2/3) = 0$$

If a game is made up of several trials, the expectation of the game is the sum of the expectations of the various trials that compose the game. It is understood that all of the trials are necessarily played.

In particular, if all of the trials are identical, the expectation of the game (composed of n trials) is equal to the product of n times the expectation of one trial. Thus if each trial is equitable, the game is equitable. As a result there is no way to make advantageous or disadvantageous, a game for which each trial is equitable.

9.6 A game with 5 silver dollars

Problem. A has 3 silver dollars and B has 2 silver dollars. The coins are all tossed, under the agreement that the player having the greatest number of heads shall win all of the 5 silver dollars, but in case of a tie, B shall win. What is the expectation of A?

Solution. There are 32 possible results in the tossing of 5 coins. In each possible result, let the first 3 positions (where capital H's and T's for heads and tails are used) represent the coins of A, and let the last 2 positions (where lower case h's and t's for heads and tails are used) represent the coins of B. The 32 possible results are

HHHhh*	HHHht*	HHHth*	HHHtt*
HHThh	HHTht*	HHTth*	HHTtt*
HTHhh	HTHht*	HTHth*	HTHtt*
HTThh	HTTht	HTTth	HTTtt*
THHhh	THHht*	THHth*	THHtt*
THThh	THTht	THTth	THTtt*
TTHhh	TTHht	TTHth	TTHtt*
TTThh	TTTht	TTTth	TTTtt

After each possible result favorable to A, we have put an asterisk. Thus the event of A winning can happen in 16 cases out of the 32, so the probability of A winning is

P (A) = 16/ 32 = 1/2

Similarly the event that B wins has probability

P(B) = 16/ 32 = 1/2

Hence player A faces the following situation

Event	Gain of A	Probability	Expectation
A wins	$2	1/2	$1.00
B wins	−$3	1/2	−$1.50

Thus the expectation of player A for the game is

$1.00 − $1.50 = −$0.50

In other words, the expectation of player A is a loss of $1.50. Thus this game is advantageous to B and disadvantageous to A.

Problem. In the above problem, what is the expectation of player A if the rules are changed so that the 2 players agree to begin again in in case of a tie?

Solution. Eliminating the ties, there are 23 possible results left. Of these, 16 are favorable to A, so his probability of winning is

$$P(A) = 16/23$$

whereas the probability of B winning is

$$P(B) = 7/23$$

The expectation of A is

$$\$2\ (16/23) - \$3\ (7/23) = (\$32 - \$21)/23 = \$11/23 = \$0.48$$

This amended game is advantageous to A and disadvantageous to B.

9.7 Betting on events.

Gambling is the act of betting on unsure events. Life and people's ingenuity provide ample events upon which people can bet. To place a bet one must first know the odds given for the event. Other names for bet are wager or stake. The bettor's odds consist of 2 numbers separated by the word "to," such as

8 to 5.

The bettor's odds represent the ratio of his potential gain to his potential loss, that is

odds = gain to loss.

For example, suppose his odds are 8 to 5 for the event, and the bettor bets 5 on the event. If the event turns out favorable, the bettor wins. His winnings are $13, made up of his $8 gain plus his $5 bet. If the event turns out unfavorable, the bettor loses. His loss is $5, made up of his $5 bet.

Another way of thinking about the bettor's odds is that the odds represent his potential gain for his bet, that is

odds = gain for bet.

For example, suppose the odds are 6 for 1 for an event, and a bettor bets $10 on the event. If the event happens, his gain is $60 for his $10 bet, making his total winnings (i.e., his gain plus his bet) $70. If the event doesn't happen, his loss is his bet of $10.

We may use either expression for odds, for example we may say the odds are "6 to 1," or we may say the odds are "6 for 1." It makes no difference. Both expressions mean the same thing.

Suppose that Boston is favored in a Boston vs. New York baseball game. An agency may give odds to bettors of 5 to 8 for Boston and 7 to 5 for New York. Thus a bettor would have to bet $8 on Boston in order to gain $5 if Boston wins. On the other hand, a bettor would have to bet $5 on New York in order to gain only $7 if Boston loses. By balancing its books (that is, by proportionately accepting bets both for Boston and for New York) the agency makes money irrespective of which team wins.

An even-money bet is one for which the odds are 1 to 1. That is, a bettor stands to gain the same sum that he bets. Suppose a bettor makes an even-money bet on an event, and his bet is $100. If the event happens, he wins $200, made up of his gain of $100 and his bet of $100. If the event doesn't happen, he loses $100, namely the amount of his bet.

Very few events in real life are considered as even money bets. Thus odds represent the equalizer that permits bets to be made on an equitable basis.

9.8 Expectation of playing roulette in Las Vegas

A roulette wheel is a precision device found in gambling halls and casinos around the world. The French roulette wheel has 37 equally spaced slots numbered

0, 1, 2, 3, 4, … , 34, 35, 36.

The zero is colored green, and half the numbers 1 through 36 are red, and the other half are black. The "French roulette wheel" is used throughout Europe.

The "Nevada roulette wheel" has 38 equally spaced slots numbered

00, 0, 1, 2, 3, 4, … , 34, 35, 36.

The zero and double-zero are colored green, and half the numbers 1 through 36 are red, and the other half are black. The Nevada wheel is usually used in those states in America where roulette is legal.

The French wheel and the Nevada wheel are the most common. However, wheels with a triple-zero (000) are not unknown in America.

For the rest of this section, we will talk about Nevada roulette such as is played in Las Vegas. One spin of the wheel represents a stochastic phenomenon. The 38 numbers (with their associated colors) represent the 38 possible results of this phenomenon. These possible results satisfy our requirement of EEE; that is, they are elemental, exhaustive, and exclusive.

In front of the roulette wheel is a table or board called the layout, or the wheel layout. The layout is divided into sections. The Nevada layout is shown in Figure 1. The red numbers and the green numbers are labeled. The unlabeled numbers are black.

		0 green		00 green
High		1 red	2 black	3 red
	1st 12	4 black	5 red	6 black
Even		7 red	8 black	9 red
		10 black	11 black	12 red
Red		13 black	14 red	15 black
	2nd 12	16 red	17 black	18 red
Black		19 black	20 black	21 red
		22 black	23 red	24 black
Odd		25 red	25 black	27 red
	3rd 12	28 red	29 black	30 red
Low		31 black	32 red	33 black
		34 red	35 black	36 red
		1st column	2nd column	3rd column

Figure 1. Nevada roulette layout

Whereas the wheel contains the possible results, the layout contains the events that you can bet on. By placing chips or money on the proper places on the layout, you are able to bet on various events.

The croupier is the man who works for the casino. His job is to spin the wheel and to pay off or collect for the casino the various bets made by the players.

The croupier spins the wheel and drops a small ball onto it while it is spinning. The players must place all bets before the ball is dropped. The ball goes in one direction and the wheel spins in the opposite direction. You sit there and wait until the ball falls into one of the numbered slots. We say that this number hits, or comes up, or shows.

If the number is favorable to the event that you bet on, then you win and you are paid by the croupier. He pays you your gain, that is, the difference between your total winnings and your bet. Your bet is left on the board ready for the next turn of the wheel. You must remove this original bet yourself if you do not want to make the same bet on the next turn. Of course many players not knowing this walk off after winning, leaving their original bet behind.

On the other hand, if the number that comes up is unfavorable to the event that you bet on, then you lose. The croupier rakes in the chips or money that you bet, and so your bet becomes your loss and the gambling house's gain.

The odds offered in gambling do not necessarily represent the true chances that the event will or will not occur. Instead the odds represent the ratio "gain to loss" on a bet. Let us give an example. Suppose that a gambler has only 5 dollars left to bet. A casino in Nevada offers odds of 2:1 (i.e., 2 to 1) that a "12-number event" will occur on one spin of the roulette wheel. The gambler puts down his stake of 5 dollars on the event. If the event occurs, the casino pays the gambler a gain of 2×5 dollars and returns his stake of 5 dollars as well. His winnings are the sum of gain plus stake. The gambler ends up with winnings of 15 dollars. It the event does not occur, the casino keeps the gambler's stake of 5 dollars. The gambler ends up with nothing.

To insure a profit on the average, the casino posts the odds. The odds include a profit margin. In our example, the event has odds 2 to 1 and the gambler puts down a stake of 5 dollars. If the event occurs, the gambler has a gain of 10 dollars, which is 2 times his stake. If the event does not occur, the gambler has a gain of −5 dollars, which is −1 times his stake. The probability on Nevada roulette of the occurrence of a "12-number event" is 12/38. Treating loss as negative gain, the expectation, or expected value, of the gambler's wager is:

$$(gain \times probability\ of\ gain) - (loss \times probability\ of\ loss)$$

which is

$$(2 \times 5 \times \frac{12}{38}) - (1 \times 5 \times \frac{26}{38}) = \frac{(120 - 130)}{38} = -\frac{10}{38}$$

This shows that the casino on the average makes 10/38, or 0.26 dollars, on every five dollar bet.

The casino chose the odds so that the payment paid to a successful gamester is less than the gain represented by the probabilities. The probability of a "12-number event" is 12/38. The favorable chances are 12. The unfavorable chances are 24+2, or 26. For a fair bet, the casino would have to post odds of 26 to 12, which is (26/12) to 1, or 2.17 to 1. The gambler puts down his stake of 5 dollars on the event. If the event occurs, the casino pays the gambler his gain of 2.17×5 dollars, which is 10.83 dollars, and returns his stake of 5 dollars as well. The gambler ends up with winnings of 15.83 dollars. It the event does not occur, the casino keeps the gambler's stake of 5 dollars. The gambler ends up with nothing.

The odds are (26/12) to 1. The stake is 5 dollars. Thus the expectation is

$$(\frac{26}{12} \times 5 \times \frac{12}{38}) - (1 \times 5 \times \frac{26}{38}) = \frac{(130 - 130)}{38}) = 0$$

If the casino posted odds on the basis of probabilities, it would be a fair game, and on the average the casino would show no profit.

Let us examine the various events that you can bet on in Nevada roulette. There are 8 types of events.

(1) The event of a number straight up

You must place your chip directly on a number. There are 38 different numbers, namely 00, 0, and 1 through 36, so there are 38 such events. Each of these single-number events carries **odds of 35 to 1**. For example, suppose that you bet 1 chip on event "number 23."

If this event happens, that is, if the number 23 comes up on the wheel, then you win. Your gain is 35 chips to 1, that is, the croupier will pay you 35 chips for the 1 chip that you bet on this event. In addition you are entitled to pick up this chip that you bet. On the other hand, if the event "number 23" doesn't happen, that is, if some other number than 23 comes up, then you lose. Your loss is the 1 chip that you bet on this event. The croupier rakes in this chip in as the house's gain.

Suppose that you bet 1 chip on event "number 23." If number 23 comes up, your **winnings are 36 chips (made up of your bet of 1 chip and your gain of 35 chips)**. If some other number comes up, your loss is 1 chip (namely, the chip that you bet).

(2) The event of 2 numbers next to each other on the layout.

You must place your chip directly on the white line between the 2 numbers. For example, to bet on the event "20 or 23", you must place your chip on the white line between 20 and 23. Each of these 2-numberevents carries **odds of 17 to 1**. Let's say that you bet 1 chip on the event "20 or 23".

If this event hits, that is, if either 20 or 25 comes up, then you win. Your gain is 17 chips to 1. The croupier will pay you 17 chips and you can pick up the, 1 chip that you bet. If this event doesn't happen, that is if any number other than 20 or 23 comes up, then you lose. The croupier rakes away the 1 chip that you placed on the white line between 20 and 23.

Suppose that you bet 2 chips on the event "number 20 or 23". If either 20 or 23 comes up, your **winnings are 36 chips (made up of your bet of 2 chips and your gain of 34 chips).** If some other number comes up, your loss is 2 chips (namely, the 2 chip that you bet).

(3) The event of 3 numbers in a layout row.

You must place your chip on the white line at the beginning of the row. For example, by placing a chip on the white line where the 7 begins, you are betting on the event "7 or 8 or 9". By placing a chip on the white line where the 31 begins, you are betting on the event "31 or 32 or 33". Each of these 3-number events carries **odds of 11 to 1**. Let's say that you bet 1 chip on the event "31 or 32 or 33".

If one of these three numbers hits, then you receive 11 chips plus the 1 chip that you bet. If some other number hits, then you lose the chip.

Hence suppose that you bet 3 chips on the event "31 or 32 or 33". If 31 or 32 or 33 hits, your **winnings are 36 chips (made up of your bet of 3 chips plus your gain of 33 chips).** If some other number comes up, your loss is 3 chips (namely, the 3 chips that you bet).

(4) The event of 4 numbers in a square on the layout

You must place your chip directly in the middle, where the 4 numbers meet each other. For example, by placing a chip on the point where 20, 21, 23, 24 meet, you are betting on the event "20 or 21 or 23 or 24". Each of these 4-number events carries **odds of 8 to 1**.

Let's suppose that you bet 1 chip on event "20 or 21 or 23 or 24". If this event happens, that is, if 20 or 21 or 23 or 24 hits, then you win. The croupier pays you 8 chips and you can pick up your bet of 1 chip. If this event doesn't happen, that is one of the favorable numbers doesn't hit, then you lose. The croupier rakes in your bet of 1 chip.

Hence suppose that you bet 4 chips on the event "20 or 21 or 23 or 24". If 20 or 21 or 2 3 or 24 hits, **your winnings are 36 chips (made up of your bet of 4 chips plus your gain of 32 chips)**. If some other number hits, your loss is 4 chips (namely, the 4 chips that you bet).

(5) The event of the 5 numbers "0 or 00 or 1 or 2 or 3".

You must place your chip on the beginning line between the single 0 and 1. This 5-number event carries **odds of 6 to 1**. Let's say that you bet 1 chip on this event. If one of the favorable numbers 00, 0, 1, 2, 3 happens, then you win and receive 6 chips plus the 1 chip that you bet. If some other number hits then you lose the chip that you bet.

Hence suppose that you bet 5 chips on the event "0 or 00 or 1 or 2 or 3". If one of these favorable numbers hits, your **winnings are 35 chips (made up of your bet of 5 chips plus your gain of 30 chips)**. If one of the other numbers bits, your loss is 5 chips (namely, your bet of 5 chips

(6) The event of 6 numbers in a rectangle on the layout.

You must place your chip on the white line at the beginning of the rectangular array of the chosen 6 numbers. For example, to bet on the event "22 or 23 or 24 or 25 or 26 or 27" you must place your chip between the 22 and 25 on the beginning line. Each of these 6-number events carries **odds of 5 to 1**. Let's say that you bet 1 chip on the event "22 or 23 or 24 or 25 or 26 or 27". If one of these numbers shows, then you receive 5 chips plus the 1 chip that you bet. If some other number happens, then you lose your bet of 1 chip.

Hence suppose that you bet 6 chips on the event "22 or 23 or 24 or 25 or 26 or 27". If one of these favorable numbers hits, your **winnings are 36 chips (made up of your bet of 6 chips plus your gain of 30 chips)**. If one of the other numbers hits, your loss is 6 chips (namely, your bet of 6 chips).

(7) The event of 12 numbers.

Underneath the numbers 34, 35, and 36 on the layout there are 3 empty boxes. These 3 boxes are called the columns, and by placing a chip in one of these you are betting on the 12 numbers in the column above. On the left of the layout there are 3 boxes, labeled 1st 12, 2nd 12, 3rd 12. The event 1st 12 happens whenever one of the numbers 1, 2, ... , 12 occurs. The event 2nd 12 happens whenever one of the numbers 13, 14, ... , 24 occurs. The event 3rd 12 happens whenever one of the numbers 25, 26, ... , 36 occurs. Each of these 12-number events carries **odds of 2 to 1**.

Let us say you bet 1 chip on the event 2nd 12. We call that betting the event 2nd dozen. If this event happens you receive 2 chips plus your bet of 1 chip. If this event doesn't happen, you lose your bet of 1 chip.

Hence suppose that you bet 12 chips on the event 2nd dozen. If one of the favorable numbers hits, your **winnings are 36 chips (made up of your bet of 12 chips plus your gain of 24 chips).** If one of the unfavorable numbers hits, your loss is 12 chips (namely, your bet of 12 chips).

(8) The event of 18 numbers.

There are 6 such events that you can bet on. They are

    (1) red: the 18 red numbers
    (2) black: the 18 black numbers

(3) odd: the 18 odd numbers 1, 3, 5, ... , 33, 35

(4) even: the 18 even numbers 2, 4, 6, ... ,34, 36

(5) low: the 18 low numbers 1, 2, 3, ... , 17, 18

(6) high: the 18 high numbers 19, 20, 21, ... , 35, 36

The location of those 6 events is on the extreme left of the layout. Each of these 18-number events carries **odds of 1 to 1**. A bet on one of these events is called an even-money bet.

Let's say that you bet 1 chip on red. If this event happens, that is, if a red number shows, then you receive 1 chip plus the chip that you bet. If this event doesn't happen, that is, if a black number or 0 or 00 shows, then you lose your bet of 1 chip.

Hence suppose that you bet 18 chips on the event red. If one of the 18 favorable numbers hits, your **winnings are 36 chips (made up of your bet of 18 chips plus your gain of 18 chips)**. If one of the 20 unfavorable numbers hits, your loss is 18 chips (namely, your bet of 18 chips).

We have now completed discussing the events that you can bet on in Nevada roulette. The following table summarizes our progress to this point.

Event	Odds	Bet	Gain	Loss
1-number	35 to 1	1	35	1
2-number	17 to 1	2	34	2
3-number	11 to 1	3	33	3
4-number	8 to 1	4	32	4
5-number	6 to 1	5	30	5
6-number	5 to 1	6	30	6
12-number	2 to 1	12	24	12
18-number	1 to 1	18	18	18

There are 38 numbers in total on the wheel. Each 1-number event has 1 chance in 38 to happen; each 2-number event has 2 chances in 38 to happen; and so on. Likewise, each 1-number event has 37 chances in 38 not to happen; each 2-number event has 36 chances in 38 not to happen, and so on. Hence we may construct the probability table:

Event	Probability to occur	Probability not to occur
1-number	1/38	37/38
2-number	2/38	36/38
3-number	3/38	35/38
4-number	4/38	34/38
5-number	5/38	33/38
6-number	6/38	32/38
12-number	12/38	26/38
18-number	18/38	20/38

If we bet 1 chip on one of each type of event, we obtain the table:

1 chip bet	Gain	Prob. of gain	Loss	Prob. of loss
1-number	35	1/38	1	37/38
2-number	17	2/38	1	36/38
3-number	11	3/38	1	35/38
4-number	8	4/38	1	34/38
5-number	6	5/38	1	33/38
6-number	5	6/38	1	32/38
12-number	2	12/38	1	26/38
18-number	1	18/38	1	20/38

Treating loss as negative gain, the expectation of an event is :

$$(gain \times probability\ of\ gain) - (loss \times probability\ of\ loss)$$

If we make use of the above table, then we obtain the following table.

Event	Expectation of betting 1 chip on event
1-number	$35/38 - 37/38 = -2/38 = -1/19$
2-number	$34/38 - 36/38 = -2/38 = -1/19$
3-number	$33/38 - 35/38 = -2/38 = -1/19$
4-number	$32/38 - 34/38 = -2/38 = -1/19$
5-number	$30/38 - 33/38 = -3/38$
6-number	$30/38 - 32/38 = -2/38 = -1/19$
12-number	$24/38 - 26/38 = -2/38 = -1/19$
18-number	$18/38 - 20/38 = -2/38 = -1/19$

We may interpret the expectation of −2/38 for betting 1 chip on an 18-number event, such as red. In 38 such bets, you expect red 18 times, you expect black 18 times and you expect green 2 times. Your gains for the 18 times that red happens cancel your losses for the 18 times that black happens. But your losses for the 2 times green happens remain. Thus you expect a loss of 2 times in 38 times, so your expectation on each spin is a loss of 2/38 of your bet of 1 chip.

From the table we see that all the roulette events that we can bet on (with 1 exception, namely the event "0 or 00 or 1 or 2 or 3") have the same expectation, namely

−2/38 per chip bet, which is −1/19 per chip bet

We may convert this fraction to percent by multiplying by 100. Thus the percentage expectation is

$(−1/19)(100) = −100/19 = −5.26$ percent.

Thus (except for one event, namely "0 or 00 or 1 or 2 or 3"), it does not make any difference upon which event you bet. The expectation will be always be a loss of 5.26 percent of each dollar you bet. In other words, the expectation for the casino is a gain 5.26 percent of each dollar that you wager.

To make the event "00 or 0 or 1 or 2 or 3" have the same expectation, the casino should give odds of 6.2 to 1 instead of 6 to 1. In other words, the odds should be 31/5 to 1. For then the expectation of this event would be

$(31/5)(5/38) − (33/38)(31/38) − (33/38) = −2/38$

which is the same as the other events that you can bet on.

Clearly the game of Nevada roulette is advantageous to the casino and disadvantageous to the player.

One method of betting popular with some gamblers at Las Vegas is the following. They will bet on red and black at the same time, but more on one than on the other. For example, on one spin on the wheel such a gambler might bet $100 on red and $200 on black. Now he must lose one of these bets regardless of what happens. If one of the green 0 or 00 hits, he loses both bets. Since his total bet is $300 and his expectation is a loss of 1/19 of each dollar bet, his expectation for his total bet is a loss of

$300 (1/19) = $15.78

The net result of his total bet is to gain $100 if black hits and to lose $100 if red hits. He could have accomplished the same result by only betting $100 on red. In this case his expectation is a loss of only

$100 (1/19) = $5.26

Thus the balancing bets of $100 on red and $100 on black puts $200 total in danger of the green 0 and 00 for no gain. Since there are 2 chances in 38 for the green to happen, his expectation from this $200 wagered is a loss of

$200 (2/38) = $10.52

which is the difference between $15.78 and $5.26.

 Of course, every roulette player has his own style of playing and his own ideas as to what events to bet on. One player used to bet on 20 to 25 single-number events on every spin. If a number that he did not bet on happened, he would say "I should have bet on that number. What bad luck!" Finally someone suggested that he should bet on all of the 38 single-number events at once. He did this, and when someone else pointed out to him that for each spin he was putting down 38 chips and getting back only 36 chips (so he was realizing his expected loss of 2 chips on each spin), he answered "I do not care. At least, I am getting a winner each time."

9.9 Expectation of playing roulette in Monte Carlo

The French roulette wheel is used in Monte Carlo. We remember that it has 37 numbers, so that the possible results of spinning the wheel are

0, 1, 2, 3, 4, ... , 34, 35, 36

The casino pays odds of 35 to 1 for single-number events. A single-number event has 1 chance in 37 to happen. Its probability to happen is 1/37, and its probability not to happen is 36/37. The expected value is also called the house average or house edge. It is the amount the player loses relative for any bet made, on average. If a player bets on a single number in the French roulette there is a probability of 1/37 that the player wins 35 times the bet, and a 36/37 chance that the player loses his bet. Thus the expectation of a bet of 1 chip placed on a single-number is

$$35 \ (1/37) - 1(36/37) = -1/37 = -0.027$$

In percentage, this expectation is

$$(-1/37)(100) = -2.70 \text{ percent}$$

Thus the expected loss of a player in Europe is 2.7 percent on bets on single-numbers.

The presence of the green 0 on the roulette wheel is the reason for the house edge. The house pays out 35 to 1 when you mathematically have a 1 out of 37 chance at winning a straight bet on a single number. To demonstrate the house edge, imagine placing straight $1 wagers on all the numbers on a roulette table (including 0) to assure a win. You would only get back 36 times your original bet having spent $37.

The six 18-number events are

Red (or rouge)	18 red numbers
Black (or noir)	18 black numbers
Odd (or impair)	18 odd numbers 1, 3, 5, ... , 33, 35
Even (or pair)	18 even numbers 2, 4, 6, ... , 34, 36
Low (or manque)	18 low numbers 1, 2, 3, ... , 17, 18
High (or passe)	18 high numbers, 19, 20, 21, ... , 35, 36

The good exceptions are the even-money bets (i.e., the six 18-number events) in some European games where the house edge is halved because only half the stake is lost when a zero comes up. Each of these six 18-number events has a payout of 1 to 1. For example, suppose you bet 1 chip on the 18-number event "red" at Monte Carlo. If red happens, then your winnings are a gain of 1 chip plus your bet of 1 chip. If black happens, then you lose the 1 chip that you bet. But if the 0 happens, then your chip is left where it stands, and the wheel is spun again. On this second spin: If red happens you take your bet of 1 chip back. If black happens the croupier takes your original chip. If the 0 happens the original chip is left on the board, and the wheel is spun for a third time, and so on.

Let us now analyze this situation. Your bet remains on the table when the 0 shows. The division as to whether you get back your bet or whether the casino takes your bet is left to further spins of the wheel. If red comes up before black, you get back your bet of 1 chip. If black comes up before red, the casino takes your bet of 1 chip. This situation is perfectly symmetrical between red and black, so the chances are equal as to whether you or the casino gets your bet of 1 chip.

Let us now suppose that you bet 1 chip on "red" for 74 spins of the wheel. Out of the 74 spins, you expect red to happen 36 times, black to happen 36 times, and the 0 to happen 2 times. Your gain from the 36 times red happens is offset by your loss from the 36 times black happens. From 1 of the 2 times that the 0 happens, you get back your bet of 1 chip, and thus neither gain nor lose. On the other time that the 0 happens, you lose your bet of 1 chip. Thus out of the 74 times that you bet on red, you expect to lose your bet of 1 chip only 1 time.

Hence your expectation is a loss of

(1 chip)(1/74)

That is, your expected loss is

1/74 = 0.0135, or 1.35 percent

of your bet by betting on "red." The same expected loss, 1.35 percent, of course holds for the five other 18 number events:

black, odd, even, low, high. This expected loss against the player on the bets on the 18-number events on French roulette is the lowest of any professional gambling game.

Let us now look at a more detailed derivation of the expected loss on the 18-number event "red" on French roulette. Let x denote the fraction

$x = 1/37$.

A player who bets 1 chip on "red" faces the following situation:

Event	Gain	Probability	Gain times Probability
Red on spin 1	1	$18x$	$18x$
Black on spin 1	−1	$18x$	$-18x$
0 on spin 1	Must spin again	x	

Red on spin 2	0	(x)(18x)	0
Black on spin 2	−1	(x)(18x)	−(x)(18x)
0 on spin 2	Must spin again	(x)(x)	
Red on spin 3	0	(x)(x)(18x)	
Black on spin 3	−1	(x)(x)(18x)	−(x)(x)(18x)
0 on spin 3	Must spin again	(x)(x)(x)	
Red on spin 4	0	(x) (x)(x)(18x)	
Black on spin 4	−1	(x) (x)(x)(18x)	−(x)(x)(x)(18x)
0 on spin 4	Must spin again	(x) (x)(x)(x)	
etc.	..	...	...
		Sum = 2	Sum = −1/74

The total expectation of the player is

$$- (x)(18x) - (x) (x) (18x) - (x) (x) (x) (18x) - ...$$
$$= -(x) (18x) [1+ x + x^2 + x^3 + ...]$$

From algebra, we know that

$$1+ x + x^2 + x^3 + ... = 1/ (1-x) = 1/ (1-(1/37)) = 1/ (36/37)) = 37/36$$

Thus the total expectation is

$$-(x) (18x) [1/ (1-x)] = (1/37) (18/37) (37/36) = -1/74 = -0.0135.$$

Thus for a $100 bet on red, the player's expectation is a loss of $1.35; that is, 1.35 percent of his bet.

Another way of finding this expectation is the following.

The probability of the player winning (W), so he gains 1 chip in addition to getting back his bet of 1 chip, is

$$P(W) = 18x = 36/74.$$

The probability of the player losing (L), so that his bet of 1 chip is taken away from him, is

$$P(L) = (18x)+ (x)(18x)+ (x) (x)(18x)+ (x)(x)(x)(18x) + ...$$
$$= (18x) [1+ (x) + (x)^2 + (x)^3+ ...]$$
$$= (18x) \{1 / [1-x)]\} = (18/37) (37/36) = 1/2 = 37/74$$

The probability of the player drawing (D), so that he gets back his bet of 1 chip but makes no gain, is

P(D) = (x) (18x) + (x) (x) (18x) + (x) (x) (x) (18x) + ...

= (x) (18x) [1 + (x) + (x)^2+ ...] = (x) (18x) [1/ (1-x)]

= (1/37) (18/37) (37/36)} = 1/74

Thus the player faces the following situation

Event	Gain	Probability	Expectation
win	1	36/74	36/74
lose	−1	37/74	−37/74
draw	0	1/74	0

His total expectation is −1/74 or −1.35 percent of his bet. French roulette is advantageous to the casino and disadvantageous to the player, but less disadvantageous than Nevada roulette.

9.10 Applications to life insurance

According to the mortality tables of an insurance company, the probability that a 25 year old man will live at least one more year is 0.993, whereas the probability that he will die within the year is 0.007. A $10,000 1-year term life insurance policy is an agreement whereby the insurance company will pay to the man's survivors $10,000 if the man dies within the year, but pays nothing if the man lives for the entire year. The company offers such a policy to a 25-year old man for a premium (that is, cost to the man) of $100.

The company faces the following situation:

Event	Gain to Company	Probability	Expectation
Man lives	$100	0.993	$99.30
Man dies	−$9900	0.007	$69.30

Hence the expectation of the company is $99.30 minus $69.30, which is $30.00 from which the company must pay costs, taxes, and dividends. If this expectation were negative, the company could not stay in business. As a game, life insurance is advantageous to the company and disadvantageous to the policy holder.

CHAPTER 10. EVENTS

Blaise Pascal wrote: We know that there is an infinite, and we know not its nature. As we know it to be false that the numbers are finite, it is therefore true that there is a numerical infinity. But we know not of what kind; it is untrue that it is even, untrue that it is odd; for the addition of a unit does not change its nature; yet it is a number, and every number is odd or even (this certainly holds of every finite number). Thus we may quite well know that there is a God without knowing what He is.

10.1 Definition of event

As we have seen, a stochastic phenomenon has different possible results. A shorter and more convenient term for possible result is "case," so we may equivalently say that a stochastic phenomenon has different cases.

For example, the spinning of a Nevada roulette wheel is a stochastic phenomenon. There are 38 possible results, or to use our alternative term, 38 cases; namely the numbers 00, 0, 1, 2, 3, ... , 34, 35, 36 each with its associated color.

We recall that the cases of a stochastic phenomenon are required to be EEE; that is,

E   elemental

E   exhaustive

E   exclusive

An event is a set of cases. Words that mean the same thing as the word "set" are "collection" and "class," so each of the following definitions of event is the same:

An event is a set of possible results.

An event is a collection of cases.

An event is a class of cases.

An event is a collection of possible results.

An event is a class of possible results.

For example, the event "odd" on Nevada roulette is the set consisting the 18 cases given by 1, 3, 5, … , 31, 33, 35. The event "even" on Nevada roulette is the set made up of the 18 cases given by 2, 4, 6, 8, … , 32, 34, 36.

The event "high" on Nevada roulette is the set of the 18 cases given by 19, 20, 21, … , 34, 35, 36, whereas the event "low" is the set of the 18 cases given by 1, 2, 3, … , 16, 17, 18.

Two important words are the words "and" and "or." The word "and" is a conjunction used to connect things that are taken jointly, like "bread and butter."

The word "or" is a conjunction used to link alternatives, like "tea or coffee." However, there are two uses of the word "or." They are called the "exclusive or" and the "inclusive or." The exclusive use may be written as "either tea or coffee" or as "either tea or coffee but not both." The inclusive use may be written as "either tea or coffee, or both" or as "tea and/or coffee." The compound word "and/or" is frowned upon, but it is useful. The expression "tea and/or coffee" indicates that tea and coffee are to be taken together or individually. It implies that either or both of the things mentioned may be affected or involved, such as "fire damage and/or water damage."

The words AND, OR, and NOT can be used as Boolean Operators. The term "Boolean searching" refers to Boolean algebra, a system of logic formulated by George Boole. In a Boolean search, keywords are combined by the operators AND, OR and NOT to narrow or broaden the search. They do not have to be entered in capitals.

The operator AND narrows the search by instructing the search engine to search for all the records containing the first keyword, then for all the records containing the second keyword, and finally show only those records that contain both keywords. If search terms are entered without an operator, then the operator AND will automatically be inserted between them.

The operator OR broadens the search to include records containing either keyword, or both. The OR search is particularly useful when there are several common synonyms for a concept, or variant spellings of a word. For example, the search "teen OR adolescent, would give all items that involve teen and/or adolescent.

The operator NOT narrows the search by excluding unwanted terms.

Use "cars AND trains" when you want to find a book that discusses both cars and trains.

Use "cars CR trains "when you want to find books that discuss cars but not trains, books that discuss trains but not cars, and books that discuss both.

Use "cars NOT trains" when you want to find books that are about cars, but EXCLUDE those books that discuss trains. For example, define A and B as A = {1, 2, 3, 4} and B = {3, 4, 5, 6}. Then we have

A AND B = {3, 4}

A OR B = {1, 2, 3, 4, 5, 6}

B NOT A = {5, 6}

A NOT B = {1, 2}

B OR (A NOT B) ={3, 4, 5, 6} OR {1, 2} = {1, 2, 3, 4, 5, 6}

B AND (A NOT B) = {3, 4, 5, 6} AND {1, 2} = { } = empty set

The empty set is the unique set having no elements. It is also denoted by the word "None" or by the symbol φ. The universal set is the set that contains all the elements. It is denoted by the word "All" or by the symbol $\mathbb{U}$.

The AND operator is used to locate records containing all of the specified search terms. For example, the search "dogs AND cats", contains all of the specified terms.

The OR operator is used to locate records matching any or all of the specified terms. For example, the search "dogs OR cats" contains either the first search term or the second. In other words, it is "dogs and/or cats."

The NOT operator is used to locate records containing the first search term but not the second. For example, the search "dogs NOT cats" contains the first search term but not the second.

The following properties of the Boolean operation AND should be noted.

(1) The Boolean operation AND has the property that "A AND B" = "B AND A."

(2) Every member of an event A is common to itself and the universal event; that is, "A AND All" = A

(3) The empty event has nothing in common with any event A; that is, "A AND None" = None.

(4) Property (3) above holds even when the event A is the empty event; that is, "None AND None" = None.

(5) Every member of an event A is common to itself; that is, "A AND A" = A.

(6) An event and its contrary event have no members in common; that is, "A AND "not A"" = None.

The following properties of the Boolean operation "OR" should be noted.

(1) The Boolean operation OR has the property that: "A OR B" = "B OR A."

(2) The members of "A OR None" are members of at least one of A, None. Because None has no members, it follows that "A OR None" = A.

(3) The members of "A OR All" include the members of All, so all cases are included. Hence "A OR All" = All.

(4) In particular, "All OR All" = All.

(5) The members of "A OR A" are just the members of A. Hence "A OR A" = A.

(6) Every case is either a member of an event or its contrary event; that is, A OR (NOT A) = All

10.2 Membership

A set is a collection of distinct and defined objects. Each object in a set S is called an element of the set. A set is represented by using braces { } with commas to separate the elements in the set. For example, the set S of all positive integers less than 5 is S = {1, 2, 3, 4}. One element of the set is the number 2. The number 3 is another element of the set. The notaion 3 ∈ S says that "3 is an element of S" or "3 is an element of S".

On the other hand, 7 is not an element of the set S. We represent this relationship as 7 ∉ S, which is read as "7 does not belong to S" or "7 is not an element of S".

A set is called finite if it contains only a finite number of elements. The set {2, 4, 6} is a finite set, as it contains only 3 elements. A set is called infinite if it contains an infinite number of elements. The set of all odd numbers is an infinite set.

Set A and set B are said to be equal sets if every element of A is in B and vice-versa. The notation for equal sets is $A = B$. In other words, two sets are equal, if $x \in A$ implies that $x \in B$ and also if $x \in B$ implies that $x \in A$. If sets A and B are not equal, then we write $A \neq B$. If A = {2, 3, 4} and if B = {2, 3, 4}, then A = B. If A = {2, 4} and B = {3}, then $A \neq B$.

On Nevada roulette, for example, the number 5 is a member of the event "odd " Likewise, the number 3 is a member of the event "low." Similarly, the number 4 is a member of the event "even," and is also a member of event "low."

We use capital letters as names of sets, for example B; and lower case letters as members of sets, for example b. There are several common ways to express membership. Each of the following sentences means the same thing:

Case b is a member of event B.

Case b belongs to event B.

Case b is in event B.

Case b is favorable to event B.

Thus for Nevada roulette all the following mean the same.

3 is a member of odd.

3 belongs to odd.

3 is in odd.

3 is favorable to odd.

Similarly, the following sentences are equivalent in expressing non-membership.

Case c is not a member of event B.

Case c does not belong to event B

Case c is not in event B.

Case c is unfavorable to event B.

Thus for Nevada roulette the following are equivalent sentences.

4 is not a member of odd.

4 does not belong to odd.

4 is not in odd.

4 is unfavorable to odd.

There are 2 common methods of specifying the membership of an event:

(1) Make a list of the members of the event. Up to now we have been using this method. For example, the event "odd" on Nevada roulette has membership given by the list:

1, 3, 5, ... , 31, 33, 35.

(2) Give a rule that specifies the members of the event. For example, the event "even" on Nevada roulette has membership given by the rule: Numbers from 1 to 36 divisible by 2.

The following sentences mean the same thing:

An event is made up of its members.

An event is composed of its members.

An event consists of its members.

An event happens when and only when 1 of its members happens.

For example, consider one spin of a Nevada roulette wheel. The event 1st-dozen happens when 1 of the numbers 1, 2, 3, ... , 11, 12 occurs, and the event 1st-dozen does not happen when one of the other numbers occurs. In other words, an event happens when 1 of its member cases occurs, and an event does not happen when 1 of its non-member cases occurs.

An event with 1 member is called a simple event. An event with more than 1 member is called a compound event. For example, the event number 5 on roulette is a simple event. The event "even" on roulette is a compound event.

It is helpful always to be mindful of the distinction between

(1) a sing e case

(2) the simple event that has this single case as its member

The distinction between a single case and the simple event made up of this single case is usually clear from the context. Cases are like gems, and events are like collections of gems. It is fundamental to the hierarchy of our thinking process to be able to distinguish between "one gem" and a "collection consisting of one gem".

10.3 Inclusion

Each of the following sentences mean the same thing; namely, that every member of event A is also a member of event B:

A is included in B

A is contained in B

A is sub-event of B

B includes A

B contains A

B is a super-event of A

On Nevada roulette, for example, the event "1st dozen" has membership 1, 2, 3, 4, ... , 11, 12 whereas the event "low" has membership 1, 2, 3, 4, ... , 17, 18. We may say any of the equivalent expressions.

Event 1st dozen is included in event low.

Event 1st dozen is contained in event low.

Event 1st dozen is a sub-event of event low.

Event low includes event 1st dozen.

Event low contains event 1st dozen

Event low is a super-event of event 1st dozen.

Not all events are comparable by this inclusion relationship. For example, event 2nd-dozen does not include event low, nor does event low include event 2nd-dozen.

Every event is included in itself; that is, "A is included in A," because the members of A (on the left) are necessarily members of the same event A (on the right).

A sub-event of an event A is called a proper sub-event of A provided the sub-event is not just A itself.

If A is included in B (i.e., every member of A is a member of B) and if B is included in A (i.e., every member of B is a member of A), then A=B (i.e., A and B have the same members).

$a \in A$	a is an element of A	set membership	A = {4,9,14}, 4 ∈ A
$A = B$	A is equal to B	both sets have the same elements	If A={3, 7, 9} and B={3, 7, 9} then A=B
A⊂B	A is a proper subset of B	All the elements of A are in B, but A has fewer elements than B	{9,10} ⊂ {9,10,25}
A ⊆ B	A is a subset of B	A is a either proper subset of B or else A = B	{9,10,25} ⊆ {9,10,25}

10.4 Universal event and empty event

A set is called a universal set if it contains every element with respect to the problem at hand. A universal set is denoted by the symbol $\mathbb{U}$ or by the word "All." The word "All" is used in the sense of "all and every," such as "all children" and "every child." The "universal event" of a stochastic phenomenon is the event made up of all the cases of the stochastic phenomenon in question.

A set is called an empty if it does not contain any element. A empty set is denoted by the symbol φ or by the word "None." An empty set is also called as a null set or void set. A empty set is also denoted by { }. The set {0} is not a null set, because this set contains the element 0. The "empty event" of a stochastic phenomenon is an empty set.

For example, for 1 spin of a Nevada roulette wheel, the universal event is the event made up of all the numbers, namely 00, 0, 1, 2, 2 35, 36.

The universal event always happens. Because the universal event always happens, it is also called the "sure event."

The "empty event" is made of none of the numbers 00, 0, 1, 2, ... , 35, 36. The empty event never happens. Because the empty event never happens, it is also called the impossible event.

For any set A, the empty set is a subset of A: $\phi \subseteq B$.

Of course, every event is a sub-event of the universal event. In other words, A is included in All, where A is any event. In particular, we see that None is included in All.

10.5 Contrary event or complement.

The relative complement of A with respect to a set B is the set of elements in B, but not in A. it is also called the relative complement of A in B, and it is denoted by $B\backslash A$. It is also denoted by $B - A$ or by B not A.

If a universe $\mathbb{U}$ is defined, then the relative complement of A in $\mathbb{U}$ is given by $A^c = \mathbb{U}\backslash A$. It is simply called the complement of A. In other words, the complement of a set A is the relative complement of A with respect to the universal set $\mathbb{U}$. The complement A^c is also denoted by A' or by "not A." If $\mathbb{U} = \{1, 2, 3, 4\}$ and A = $\{1, 3\}$, then $A^c = \{2, 4\}$.

Two people are contrary if they do everything the opposite. Likewise, for every event there is a contrary event. If we let A designate an event, then we shall let "not A" designate the contrary event. An alternate term for the contrary event is complement, with the symbol A^c.

In set theory, the complement of a set A is made up of the elements not in the set A. In other words, given a set A, the complement of A is the set of all elements in the universal set $\mathbb{U}$, but not in A.

If we remove the members of A from the universal event, then we are left with the non-members of A. These non-members of A make up the contrary event, designated either by "not A" or by A^c. This usage of "not" is its usage as an operation of Boolean algebra. The non-members of A make up the event "not A." The non-members of "not A" make up the event A. The 2 events A and "not A" have no cases in common.

Everything in the universal event is either a member of A or a member of "not A."

For example, on 1 spin of Nevada roulette, the event odd is made up of the numbers

1, 3, 5, 7, ... , 33, 35.

The contrary event, not odd, is made up of the numbers

00, 0, 2, 4, 6, ... , 34, 36.

The 2 events "odd" and "not odd" have no numbers in common, and all the numbers belong to one or the other of these 2 events.

The contrary event to the universal event is the empty event; that is,

"not All" = None.

The contrary event to the empty event is the universal event; that is,

"not None" = All.

In summary, the complement of the universal set is the empty set. $\mathbb{U}^c = \emptyset$. The complement of the empty set is the universal set; $\emptyset^c = \mathbb{U}$.

Because A is made up of its members, "not A" is made up of the non-members of A. It follows that "not (not A)" is made up of the members of A. Therefore the contrary of the contrary of an event is the event; that is, "not not A" = A." In symbols, we have $(A^c)^c = A$

In summary, the complement of a set is a set that contains the elements that are contained in the universal set, but not in the given set. The complement of A consists of elements that are not in A but are in $\mathbb{U}$.

| A^c | complement of A | all the objects that do not belong to set A | If $\mathbb{U}$ = {1, 2, 3, 4} and A = {1, 3}, then A^c= {2, 4} |
| A \ B | relative complement | objects that belong to A and not to B | If A = {4,9,14} and B = {1,2,4}, then A \ B = {9,14} |

10.6 Common event or intersection

Two people can have things in common. Likewise 2 events can have members in common. For example, on a spin of Nevada roulette, the events low and 2nd-dozen have 6 numbers in common. We recall that low is made up of the numbers

1, 2, 3, 4, 5, 6, 7, 8, 9, 10, 11, 12, 13, 14, 15, 16, 17, 18,

and that 2nd-dozen is made up of the numbers

13, 14, 15, 16, 17, 18, 19, 20, 21, 22, 23, 24.

Hence these 2 events have the numbers

13, 14, 15, 16, 17, 16, 19

in common. If any one of these 6 numbers happens, then both the event low and the event 2nd-dozen happen. For this reason, let us designate the event made up of these 6 numbers by "low and 2nd-dozen."

Given any 2 events A, B we can form their "common event" or "intersection." The intersection is defined to be the event made up of the common members of the separate events A, B. The common members are the ones that are members of both the event A and the event B. The intersection of two or more sets is the set containing only the common elements among all the sets under consideration. The intersection of A and B is written as "A and B." Another notation for "A and B" is $A \cap B$, which is read A intersection B. In other words, the intersection operation is denoted either by the symbol $\cap$ or by the word "and". If A = {1, 2, 3, 4, 5} and B = {1, 3, 5, 7, 9}, then $A \cap B$ = {1, 3, 5}. Whenever you see the word "and" in quotation marks, it is a sign-post to tell you that "and" is used in this specialized sense.

In summary, the intersection of the sets A and B may be written as $A \cap B$ or as "A and B". The intersection of two sets A and B is the set that contains all elements of A that also belong to B (or equivalently, all elements of B that also belong to A), but no other elements. We have:

| $A \cap B$ | intersection | Objects that belong to both set A and set B | If A= {2, 3, 4} and B= {3, 4, 5}, then $A \cap B$ ={3, 4} |

Problem 1. Suppose A has members a, b, c, d and suppose B has members e, f, g. What is the membership of "A and B"?

Solution. The event A has no members in common with the event B, so their common event is empty; that is, "A and B" = None.

Problem 2. Suppose A has members a, b, c, d and suppose B has members c, d, e, f. What is "A and B"?

Solution. The common event A and B is the event with the common members of A, B; namely, the members c, d.

10.7 Incompatible events

Two people are incompatible if they have nothing in common. Likewise, 2 events are "incompatible" provided they have no members in common. In other words, 2 events are incompatible provided that they cannot both happen at the same time.

Alternative names for "incompatible events" are "disjoint events" or "mutually exclusive events." We recall that the common event "A and B" is made up of the members that are common to the events A, B. Thus events A, B are incompatible if and only if their common event is the empty event, that is, "A and B" = None.

For example, on Nevada roulette, the events "even", "odd" are incompatible. They have no common numbers. They cannot both happen at the same time. An event and its contrary event are incompatible; that is, "A and not A" = None.

Because a simple event has only 1 member, it follows that each simple event is incompatible with every other simple event.

10.8 United event or union

People who play bridge can unite with people who play poker to form a united card playing organization. The united organization is made up of people who play bridge or poker. But suppose a person plays both bridge and poker. Certainly he will be a member of the united organization, but he will be counted only as 1 member, not as 2 members. Thus he has no preferential status with respect to a person who plays bridge but not poker, who is also counted as 1 member of the united organization. Likewise a person who plays poker but not bridge is counted as 1 member of the united organization.

For example, if in town there are:

800 people who play bridge but not poker,

200 people who play bridge and poker,

500 people who play poker but not bridge,

then the united organization will have

800 + 200 + 500 = 1500 members.

Another way of obtaining the number of members in the united organization is by use of the data:

1000 people play bridge (regardless of whether or not they play poker),

700 people play poker (regardless of whether or not they play bridge),

200 people play both bridge and poker.

Then the united organization has

1000 + 700 − 200 = 1500 members,

which agrees with the above number of members.

Given any 2 events A, B, we can form their united event. The united event, designated by "A or B," is called the union of A, B. Its members are members of at least one of the events A, B. Whenever you see the word "or" in italics, it should serve as a sign-post to tell you that "or" is used in this specialized sense. The union is denoted either by the word "or" or by the symbol ∪. We can write either "A or B" or A ∪ B, which is read A union B.

The union of a collection of sets is the set of all distinct elements in the collection. The union of two sets A and B is the collection of points which are (1) in A not B, (2) in B not A, (3) in both A and B The union of A = {1, 2, 3} and B ={2, 3, 4} is A ∪ B -= {1, 2, 3, 4}. Note that the elements 2, 3 belong to A and also to B. Since sets cannot have duplicate elements, the union contains 2, 3 but not 2, 3, 2, 3.

Another example is afforded by the set of even numbers {2, 4, 6, 8, 10, ...}, and the set of prime numbers {2, 3, 5, 7, 11, ...}. The number 9 is not contained in their union because 9 is neither prime nor even.

The union is a fundamental operation through which sets can be combined and related to each other. The union of a collection of sets is the set of all distinct elements in the collection. The union of two sets A and B is the collection of points which are "only in A" or "only in B" or "in both A and B." If x represents a single element, then x is a member of the union if it is an element present in set A or in set B, or in both.

In summary,

| A ∪ B | union | Objects that belong to only set A, to only set B, and to both A and B | If A= {2, 3, 4} and B= {3, 4, 5}, then A ∪ B ={2, 3, 4, 5} |

Problem 3. Suppose A has members a, b, c, d. Further suppose B has members e, f, g. What is the membership of" A or B"?

Solution. The united event A ∪ B is made up of all members of at least one of A, B. Hence A or B has membership a, b, c, d, e, f, g.

Problem 4. Suppose A has members a, b, c, d. Further suppose B has members c, d, e, f, g. What is "A or B"

Solution. The united event A or B is made up of all members of at least one of A, B. Hence or B has membership a, b, c, d, e, f, g. We note that although c, d are both members of A and members of B, they have no preferential status in "A or B."

10.9 De Morgan's laws

The two complement laws are:

The complement of the universal set is $\mathbb{U}^c = \emptyset$.

The complement of a null set is $\emptyset^c = \mathbb{U}$.

The two identity laws are:

The union of A with the empty set is $A \cup \phi = A$

The intersection of A with the universal set is $A \cap \mathbb{U} = A$

The two domination laws are:

The union of A with the universal set is $A \cup \mathbb{U} = \mathbb{U}$

The intersection of A with the empty set is $A \cap \phi = \phi$

Three statements of De Morgan's first law are

The complement of a union is the intersection of the complements
$(A \cup B)^c = A^c \cap B^c$
"not (A or B)" is the same as "(not A) and (not B)"

Three statements of De Morgan's second law is

The complement of an intersection is the union of the complements
$(A \cap B)^c = A^c \cup B^c$
"not (A and B)" is the same as "(not A) or (not B)"

For example, let $\mathbb{U} = \{1, 2, 3, 4\}$, A = {2, 3}, B = {3, 4}. Then the two De Morgan laws may be illustrated by the table:

$\mathbb{U}$	1	2	3	4	
A		2	3		
B			3	4	
A^c	1			4	
B^c	1	2			
$A \cup B$		2	3	4	
$(A \cup B)^c$	1				First law
$A^c \cap B^c$	1				$(A \cup B)^c = A^c \cap B^c$
$A \cap B$			3		
$(A \cap B)^c$	1	2		4	Second law
$A^c \cup B^c$	1	2		4	$(A \cap B)^c = A^c \cup B^c$

CHAPTER 11. FAVORABLE CHANCES TO TOTAL CHANCES

Blaise Pascal wrote: Knowledge of physical science will not console me for ignorance of morality in time of affliction, but knowledge of morality will always console me for ignorance of physical science.

11.1 Phenomena whose cases are called chances

Now we want to look at stochastic phenomena that have a finite number of possible results. Thus we may list all the cases for such a phenomenon.

For example, in tossing a coin there are 2 cases, namely H, T. In tossing a die there are 6 cases, namely 1, 2, 3, 4, 5, 6. For example, in picking a ball from an urn containing 3 balls colored red, white, blue there are 3 cases, namely the red ball, the white ball, and the blue ball.

For many such finite phenomena, there exists a symmetry among the cases. For example, a die is in the form of a cube and is manufactured from a homogeneous material. Thus its 6 faces have symmetry that tells us that the turning up of one side should not be more propitious than the turning up of any other side.

If we put into an urn some balls of the same weight and same material, differing only in color, and if we put our hand into the urn and take out a ball without looking, there is a symmetry here that tells us that the drawing of one ball should not be more propitious than the drawing of any other ball.

We must use our judgment to decide for each particular phenomenon whether its cases do or do not have the necessary kind of symmetry. In other words, for the phenomenon in question we must decide whether or not any case is more propitiously endowed than any other case. To take this decision we must make use of our accumulated experience as well as theoretical reasoning by analogy.

When the cases of a stochastic phenomenon do have this kind of symmetry, that is, when no case is more propitious than any other case, then each case is called a chance. In other words, the cases of such phenomena are called chances. To repeat, the possible results of such phenomena arc called chances.

For example, when we toss a homogeneous, symmetric coin there are 2 chances, namely H, T. When we toss a homogeneous symmetric die, there are 6 chances, namely 1, 2, 3, 4, 5, 6. When we draw 1 ball from an urn containing 3 l ke balls colored R, W, B, there are 3 chances, namely R, W, B.

This usage of the word chance is commonplace and well accepted in the English language. In fact, this usage is so basic and so in accord with experience and analogy that it takes its place as a primitive notion.

11.2 Probability

Let us now define probability for events associated with those phenomena whose possible results are chances. We recall that an event is a set of possible results. In the present context, an event is a set of chances. For example, for each turn of a Nevada roulette wheel, there are 38 chances, namely

00, 0, 1, 2, 3, 4, ... , 34, 35, 36.

The event low is made up of 10 chances, namely

1, 2, 3, 4, ... , 16, 17, 18.

These 18 chances are the favorable ones for the event low. The event low happens when and only when one of these 18 favorable chances happens. Thus we may say any of the following sentences, all of which mean the same thing.

(1) The event low has 18 favorable chances out of 38 total chances to happen.

(2) Low has 18 favorable out of 30 total chances to happen

(3) Low has 18 chances in 38.

(4) The chances that the event low happens are 18 in 38.

(5) The chances of low are 18 in 38.

From the statement about the chances of low, it is an easy step to go to the equivalent statement about the probability of low. All of the following sentences mean the same thing.

(1) The event low has a probability of 18/38 to happen.

(2) Low has a probability of 18/38 to happen.

(3) Low has a probability of 18/38.

(4) The probability that the event low happens is 18/38.

(5) The probability of low is 18/38.

In mathematical symbols, we write P(low)= 18/38, which is read as probability of low equals 18/38.

Using the above example as a guide we therefore introduce the following definition of probability as a ratio of chances. Probability is the ratio of the number of favorable chances to the number of all chances. In other words, the probability of an event is the ratio of number of chances favorable to the event to the total number of chances. As a guide to memory, we write

probability = favorable chances / total-chances

In still other words, if an event has n chances in N, then its probability is the ratio n/N.

For example, when we toss a good coin, H has 1 chance in 2, so the probability of H is 1/2. When we toss a good die, the event 5 (i.e. the face 5 lands up) has 1 chance in 6, so its probability is 1/6. When we say that the event of each face on a good die has a probability of 1/6, we always must take into consideration that the manufacture of a so-called "good" die can only be realized in practice to a certain approximation. The same situation holds for all the questions studied by means of probability theory.

The difference between a mathematical model and a mathematical theory should be noted. A mathematical theory does not need any connection whatsoever with the real world. A mathematical model is a mathematical theory that can be applied to real phenomena. The definition of probability given here represents a mathematical model. There are many applications of this model to real phenomena. In fact this probability model was discovered in the Renaissance by the empirical study of some real phenomena, namely the tossing of dice and the dealing of carols in games of chance.

Gambling is the largest industry in the United States, in dollar amounts larger than steel, automobiles, or oil, and more money is wagered on dice tossing than all other games of chance combined. Because the rules of games of chance are based on the above definition of probability, the applicability of this model to such phenomena cannot be denied. Nevertheless, an important point to remember is that this model also is applicable to many phenomena occurring in science and technology.

The intensity of gambling throughout the ages had been so great that the relative probabilities of the various throws of dice and various card hands were known empirically by the time of the Renaissance. From this empirical knowledge came the abstraction to the theoretical concept of probability as the ratio of favorable chances to total chances. This definition was used throughout the first extant book on probability theory; namely the book *Libor de Ludo Aleae* by Jerome Cardano, written about 1520, one hundred years before the Pilgrims landed at Plymouth.

Of course, we must always bear in mind that the application of this definition is limited to those phenomena whose eases may be considered as chances.

Our plan of attack in this book is first to make firm the ideas behind this limited definition, and then to proceed to the more general definition of probability.Hence for the next several chapters, we will concentrate for the most part on the definition of probability as the ratio of the number of favorable chances to the number of all chances.

11.3 Essential properties of probability

We can let N denote the total number of chances, and let n denote the number of chances favorable to an event. Let us call this event E. The probability of the event is given by the formula

$P(E) = n/N$

We see that the probability is a fraction. This fraction is equal to 1 if and only if n=N. But n=N if and only if all the chances are favorable for the event in question. That is, the event in question consists of all the chances (i.e., the event in question is the universal event). Thus the probability of en event is 1 if and only if the event is sure to happen.

This fraction is equal to 0 if and only if n=0. But n=0 if and only if no chances are favorable. That is, the event in question consists of no chances (i.e. the event is the empty event). Thus the probability of an event is 0 if and only if the event is sure not to happen.

The probability of any other events lies between these 2 extremes, namely between 0 and 1.

Whenever we consider the probability of an event, we may also consider the probability of the contrary event. If the event is denoted by E, then the contrary event is denoted by "not E." All the chances, except the chances favorable to the event E, make up the contrary event. For example, a urn contains N balls among which n are white and the others are red, yellow, black. Let the event E be the drawing of a white ball. The contrary event "not E" is the drawing of a ball that is not white. Since the event E is made up of n favorable chances (since there are n white balls) its probability is

   $P(E) = n/N$

Since the event mot E is made up of N-n favorable chances (since there are N-n red, yellow, and black balls), its probability is

   $P(not\ E) = (N-n) / N = 1 - (n/N)$

Let us now add these 2 probabilities together. We see that

   $P(E) + P(not\ E) = 1.$

In words: The sum of the probabilities of an event and the contrary event is equal to 1.

Problem 1. If the probability of an event happening is 1/5, what is the probability of its not happening?

Solution. 1 - 1/5 = 4/5.

Problem 2. If the probability of a shot hitting a target is 81/100, what is the probability of its not hitting?

Solution. 1 – 81/100 = 19/100.

Problem 3. If the probability of a student passing an examination is 70/100, what is the probability of his not passing?

Solution. 1 - 70 /100 = 30/100.

Problem 4. If the probability of an experiment succeeding is 5/9 what is the probability of its failing?

Solution. 1 – 5/9 = 4/9.

Problem 5. If the probability of A winning a certain race is 1/5 and the probability of B winning is 1/6, what is the probability that neither should win?

Solution. A has 1 chance in 5 to win, or equivalently he has 6 chances in 30. B has 1 chance in 6 to win, or equivalently he has 5 chances in 30. Hence there is 5 + 6, which is 11, chances in 30 that one of them should win; that is, a probability of 11/30. Therefore the probability that neither should win is 1 – 11/30, which is 19/30.

We must carefully note that the contrary event is the collection of all those chances that are not favorable to the event. In other words the dichotomy must be complete between the event and the contrary event. For example, consider a game of that may result in win, lose, or draw. Let the event be win. The contrary event would include both lose and draw, not just lose.

We say that an event is very probable if its probability is very close to 1. The probability of the contrary event would be close to 0. In ordinary language we often say that the probability of the event is very great. Of course, this manner of speaking is not in accord with the definition of probability which does not permit the probability to be larger than 1. Nevertheless, the usual usage has the following explanation. One compares in his mind the probability of an event to the probability of the contrary event. If the probability of an event is close to 1, then probability of the contrary event is close to 0. Thus the ratio of the 2 probabilities will be great.

For example, if an urn contains 1000 balls, 999 of which are white, we would say in ordinary language that the probability is great that we would draw a white ball. The probability of this event is 999/1000. The probability of the contrary event is 1/ 1000. We mean that the ratio of these 2 probabilities, namely

(999/ 1000) / (1/1000) = 999

is great. We may call the ratio the relative probability. In conclusion, we see that in ordinary speech we often have in mind relative probabilities instead of probabilities. Whenever the probability of en event is very close to 1, its relative probability is very large.

11.4 Problems based on chances

There are many problems based on the definition of probability as the ratio of favorable chances to total chances. These problems are interesting in their own right and instructive in that they provide background for the more abstract formulation of probability that we will come to later.

Problem 6. A friend sets out with 9 other passengers on a ship that had a crew of 20 men. News comes that a man fell overboard. What is the probability that he is our friend?

Solution. The total number of men on the ship was 30. Hence there is 1 chance in 30, or a probability of 1/30, that our friend fell overboard.

Problem 7. (Continuation of Problem 6). Suppose that further news comes that it was a passenger that fell overboard. What is the probability that he is our friend?

Solution. The total number of passengers on the ship was 10. Our friend was 1 of these 10 passengers. Hence there is 1 chance in 10, or a probability of 1/10, that our friend fell overboard.

Problem 8. (Continuation of Problem 7). What is the probability that the man overboard is not our friend?

Solution. There are 2 alternative solutions to this problem. One solution is obtained by noting that the event "the man overboard is not our friend" is the contrary to the event "the man overboard is our friend", which by problem 2 is 1/10. Hence the desired probability is

   1 — 1/10 = 9/10.

The other solution is obtained directly from the definition of probability. There are 9 chances in 10 that the man overboard is not our friend, so the probability is 9/10.

Problem 9. A party of 10 young people is seated at a round table. John and Elizabeth are in this party, and they hope to sit next to each other. What is the probability that they will sit together?

Solution. Excluding John's place, whatever it might be, there are 9 places. Of these 9 places, 2 are adjacent to John and 7 are not adjacent. Thus Elizabeth has 2 chances in 9 to sit next to John. Thus the desired probability is 2/9.

Problem 10. The four letters e, n, t, s are put in a row in any order. That is the probability that they form an English word.

Solution. There are 24 possible orders of these 4 letters; namely,

ents	(nets)	(tens)	(sent)
enst	(nest)	tesn	setn
etns	ntes	tnes	snet
etsn	ntse	tnse	snte
esnt	nsnt	tsen	sten
estn	nstn	tsne	stne

Out of these 24 chances, the 4 in parenthesis are favorable to the event. Hence the probability is 4/24, or 1/6.

Problem 11. If a letter be taken from the word "association," what is the probability that it is a vowel?

Solution. The word associations has 11 letters, out of which 6 are vowels (namely a, o, c, i, a, i, o). Hence the desired probability is 6/11.

11.5 Conclusions

In this chapter, we introduced the definition of probability based on chances: The probability of an event is the ratio of the number of favorable chances to the number of chances. Thus, on the basis of this definition, the probability would be 2/7 for selecting haphazardly a red ball from a box containing 2 red balls and 5 black balls. This probability is based on the argument that there are 7 possible results of the haphazard selection, corresponding to the 7 balls in the box, and that 2 of them are favorable to the event in question.

Nevertheless it is easy to show that this definition, as it stands, can lead to different evaluations of the probability of an event. What is the probability of getting 2 heads (i.e., HH) in 2 tosses of a coin? One can argue that this probability is 1/4 because there are 4 possible results, namely

HH, HT, TH, TT.

One can also argue that this probability is 1/3 because there are 3 possible results, namely 2 heads, 1 head, or no heads, that is,

2H, 1H, 0H.

This difficulty of more than 1 evaluation of a given probability arises from the fact that alternatives of unequal range have been treated as chances. Thus the case of throwing 1H in 2 tosses can be disjunctively divided into

HT, TH.

Thus in using this definition, it is required that each case be "equally probable." Another term that means the same is "equally likely." Alternatively, we may say that each case has an "equal probability" or else an "equal likelihood." Superficially it appears that the requirement that the cases be equally probable makes circular the definition of probability based on chances. In other words, it looks as though we were defining probability in terms of equal probability. Therefore, this definition is only applicable to those phenomena where equal probability can be defined without reference to probability. Such is done when we admit the possibility of a perfect die thrown perfectly, or a perfect roulette wheel spun perfectly, or a perfect pack of cards shuffled perfectly. For then, the circularity is an innocuous verbal circularity, which is logically harmless. In conclusion, under such perfect physical conditions, the definition of probability as the ratio of favorable chances to total chances is not circular.

> Blaise Pascal wrote: It is not from space that I must seek my dignity, but from the government of my thought. I shall have no more if I possess worlds. By space the universe encompasses and swallows me up like an atom; by thought I comprehend the world.

CHAPTER 12. ADDITION OF PROBABILITIES

Blaise Pascal wrote: When we wish to demonstrate a general theorem, we must give the rule as applied to a particular case; but if we wish to demonstrate a particular case, we must begin with the general rule. For we always find the thing obscure which we wish to prove, and that clear which we use for the proof; for, when a thing is put forward to be proved, we first fill ourselves with the imagination that it is therefore obscure, and on the contrary that what is to prove it, is clear, and so we understand it easily.

12.1 Addition of probabilities for incompatible events

One of the most important, and also the simplest, rules of probability is the addition rule. We have already made use of this rule several times. For example, on Nevada roulette there are 18 chances in 38 for red and 18 chances in 38 for black. The events red and black are incompatible. Either red can happen or black can happen, but not both. Thus the united event "red or black" has 18 + 18 = 36 chances in 38 to happen. Thus the sum of the probability that red happens plus the probability that black happens gives the probability that red or black happens as

$(18/38) + (18/38) = 36/38$

We may summarize our reasoning as follows:

(1) We verify that 2 events are incompatible

(2) We find the probability of their united event by adding their separate probabilities.

12.2 The addition rule

The fundamental rule upon which the rule for the addition of probabilities rests is the addition rule.

ADDITION RULE: If one thing can be done in m different ways and another thing can be done in n other different ways, then either one thing or the other can be done in the sum of m plus n different ways.

This rule is basic, and goes back to when you first learned how to add. For example, if we put 5 apples and 6 oranges into a box, then the number of apples or oranges in the box is 5 + 6, or 11.

As another example, if John knows 4 kinds of trees and if Mary knows 3 other kinds, then John or Mary know 4 + 3 = 7 different kinds. Suppose John knows oak, elm, maple and birch, and Mary knows cedar, pine, and spruce. Then oak, elm, maple, birch, cedar, pine and spruce are known by John or Mary.

However, if John knows 4 kinds of trees and if Jane knows 3 kinds, then we cannot conclude that John or Jane know 4 + 3, or 7, different kinds. The reason is that Jane might not know 3 other kinds, but instead knows some of the same kinds as John. Suppose now that Jane knows oak, elm and pine. Then oak, elm, maple, birch, and pine are known by John or Jane.

Thus the word "other" is an essential part of the addition rule.

A set that has only a finite number of members is called a finite set. Let this finite number be denoted by n. The number n is a positive whole number, such as 1, or 2, or 5, or 5000, or 500,000. We exclude n=0, which would correspond to the empty set. Instead of saying "a set with n members," it is much more convenient to use a shorter expression. Hence let us agree that the expression "n-set" means "a set with n members."

Now we can state the addition rule in terms of sets.

> RULE: Let A be an m-set and B be an n-set. Let A and B have no members in common, that is the common set (or intersection) satisfies "A and B" = None. Then the united set (or union) "A or B" is an (m + n)-set.

12.3 Review

Given a stochastic phenomenon, we recall that:

(1) Its possible results are called cases.

(2) The cases are required to be EEE, that is, elemental, exhaustive, and exclusive.

(3) An event is a collection of cases.

(4) 2 events are called incompatible provided the 2 events have no cases in common.

There is an important kind of stochastic phenomenon that has the 2 properties:

(1) It has only a finite number of cases

(2) No case is more propitiously endowed than any other case. Each case of this kind of phenomenon is called a chance. For the time being, we are only dealing with this type of phenomenon.

The number of chances for the entire stochastic phenomenon is called the total number of chances. An event is a collection of chances. The number of chances belonging to the event is called the number of favorable chances. The probability of the event is the ratio of the number of favorable chances to the total number of chances.

12.4 Rule for the addition of 2 probabilities

Suppose we have an urn that holds 20 like balls, among which:

5 are colored amber (A)

7 are colored blue (B)

8 are other colors.

The probability of extracting an amber ball is

$P(A) = 5/20.$

The probability of extracting a blue ball is

$P(B) = 7/20.$

Then the probability of extracting either an amber or blue ball is evidently

$P(A \text{ or } B) = 12/20.$

One sees immediately the equation

$(5/20) + (7/20) = 12/20$

which is

$P(A) + P(B) = P(A \text{ or } B).$

This equation illustrates the rule for the addition of probabilities. In order to understand this rule, it is important to know precisely the conditions under which we obtained this equation.

The desired event is the event of extracting an amber or blue ball. This event is denoted by

"A or B."

There are 2 kinds of favorable chances for this event. One kind is made up of the chances for the extraction of an amber ball, that is, the chances for the event A. The other kind is made up of the chances for the extraction of a blue ball, that is, the chances for the event B.

The event A has no chances in common with the event B, because if the extracted ball is amber then it cannot be blue, and if the extracted ball is blue then it cannot be amber. Thus the 2 events A, B are incompatible. The event A has 5 chances in 20. The event B has 7 chances in 20. The 5 chances of A are distinct from the 7 chances of B. These 2 events cannot both happen at the same time. Because the events A, B have no chances in common, the number of chances for the event A or B is the sum of the chances for A plus the chances for B. That is, the number of chances for the event A or B is 5 + 7 = 12. Because the chances add, the probabilities also add. Hence the probability of the event A or B is the sum of the probability of A plus the probability of B. In other words, the probability of the event A or B is

$(5/20) + (7/20) = 12/20$

This equation illustrates the rule for the addition of probabilities.

We can embody this rule in the following statements. We are given that:

(1) The phenomenon has N chances in total.

(2) Event A has a favorable chances.

(3) Event B has b favorable chances.

(4) Events A, B are incompatible. (In other words, the a chances of A are different from the b chances of B).

Then it follows that:

(1) The united event A or B has (a + b) favorable chances.

(2) Thus the probability of event A or B is $P(A \text{ or } B) = (a + b)/N$

(3) This equation may be written $P(A \text{ or } B) = (a/N) + (b/N)$.

(4) But the probability of the event A is $P(A) = a/N$

(5) Also the probability of the event B is $P(B) = b/N$

(6) Hence the probability of the united event is $P(A \text{ or } B) = P(A) + P(B)$.

This result may be stated as follows:

> RULE: If the events A, B are incompatible, then the probability of the united event "A or B" is equal to the sum of the probability of A plus the probability of B.

Also the rule may be stated more concisely in symbolic form as follows:

> RULE: If the common event "A and B" (or intersection A∩B) is Null, then the united event "A or B" (or union $A \cup B$) has probability P(A or B) = P(A) + P(B).

The rule states, simply, that the probabilities of 2 incompatible events can be added. The sum of the probabilities is the probability that one or the other of these 2 events happens.

12.5 Some problems and solutions

In the application of this theorem, it is very important to verify that the condition of incompatibility is fulfilled.

Problem 1. Suppose that there are 30 airplanes at the London airport that will fly within the hour. Of these 30 airplanes, 5 are going west to New York and 3 others are going east to Moscow. The New York set and the Moscow sets of airplanes are incompatible. Thus the number of airplanes going to either New York or Moscow is 5 + 3 = 8. You arbitrarily get on one of the 30 airplanes. What is the probability of going to New York or Moscow?

Solution. You have 5 chances in 30 to go to New York, and 3 chances in 30 to go to Moscow. The two destinations are incompatible. Thus you have 8 chances in 30 to go to New York or Moscow, and so the probability of this event is 8/30.

Now let us suppose that 1 of the 5 airplanes going to New York includes a stop at Iceland. Also suppose there is no other of the 30 airplanes going to Iceland. The number of airplanes going to New York or Iceland is not 6, but 5. In other words, the airplane to Iceland is the same as one of the airplanes to New York, and so the event of going to Iceland is not incompatible with the event of going to New York. Your probability of going to Iceland or New York is 5/30, not 6/30.

Problem 2. A coin is tossed. That is the probability of T (i.e. tails)?

Solution. The tossing of 1 coin has 2 possible results, namely H, T. Thus there is 1 chance in 2 for the event of T to happen, so the desired probability is 1/2.

Problem 3. 2 coins are tossed. What is the probability of at least 1T (i.e. at least one tails)?

Solution. The tossing of 2 coins has 4 possible results; namely, HH, HT, TH, TT. The event of 1T (i.e. 1 tails shows) has 2 chances; namely, HT, TH. The event of 2T (i.e. 2 tails show) has 1 chance; namely, TT. The event of 1T is incompatible with the event of 2T. The event of at least 1T is the same as the event of 1T or 2T. Hence this event has 3 chances; namely, HT, TH, TT. Hence the desired probability is 3/4. In other words,

P(at least 1T) =P(1T or 2T) = P(1T) + P(2T) = (2/4)+(1/4) = 3/4

Problem 4. 3 coins are tossed. What is the probability of at least 1T?

Solution. The tossing of 3 coins has 8 chances in total; namely,

HHH, HHT, HTH, THH, HTT, THT, TTH, TTT.

The event 1T (i.e. 1 tails shows) has 3 chances; namely, HHT, HTH, THH. The event 2T (i.e. 2 tails show) has 3 chances; namely, HTT, THT, TTH. The event 3T (i.e. 3 tails show) has 1 chance; namely, TTT. The event of at least 1T is the union of the events 1T, 2T, 3T; that is, at least 1T = 1T or 2T or 3T.

The events 1T, 2T, 3T are incompatible. Hence the event of at least 1T has 3 + 3+ 1 = 7 chances, so its probability is 7/8. Because P(1T) = 3/8, P(2T) = 5/8, P(3T) = 3/8, we see that

7/8 = (3/8) + (3/8) + (1/8)

or

P(1T or 2T or 3T) = P(1T) + P(2T) + P(3T).

The equation represents the theorem of total probability for 3 incompatible events.

Problem 5. 2 dice are thrown. What is the probability that they show 7 or 11?

Solution. There are 36 possible results by tossing 2 dices. These 36 chances are

11	12	13	14	15	(16)
21	22	23	24	(25)	26
31	32	33	(34)	35	36
41	42	(43)	44	45	46
51	(52)	53	54	55	[56]
(61)	62	63	64	[65]	66

For each case, the first number is the face of one die and the second number is the face of the second die. The chances in parentheses are the ones that give 7 and the chances in brackets are the ones that give 11. The event of 7 is incompatible with event of 11. Since the probability of 7 is P(7) = 6/36 and the probability of 11 is P(11) = 2/36. Hence the probability of 7 or 11 is

P(7 or 11) = (6/36) + (2/36) = 8/36 = 2/9 = 0.2222

Problem 6. A card is drawn from a well-shuffled pack of 52 cards. What is the probability of getting a king or a queen?

Solution. If the extracted card is a king, then it is not a queen. If it is a queen, then t is not a king. Hence, the 2 events are incompatible. There are 4 chances in 52 for a king, so the probability of a king is 4/52. Likewise there are 4 chances in 52 for a queen, so the probability of a queen is 4/52. Hence the probability of a king or queen is

(4/52) + (4/52) = 8/52.

Problem 7. Ann and Bob toss a coin under the following conditions. If toss 1 shows H, Ann wins. If toss 1 shows T, however, the coin must be tossed 2 more times. Then, if out of the 3 tosses, H shows at least 2 times, Ann also wins. What is the probability that Ann wins?

Solution. We could reason as follows. Ann wins in 2 different ways. The first way is for H to come up on toss 1. The probability of this event is 1/2. The second way is for H to come up at least 2 times out of 3 tosses. The probability of this event is also 1/2. Thus the probability that Ann wins is the sum of these 2 probabilities. This sum is

(1/2) + (1/2) = 1

which says that it is certain that Ann wins. But this is absurd.

Instead, we must consider the set of all possible outcomes, namely

HHH, HHT, HTH, THH, HTT, THT, TTH, TTT.

Now the chances favorable to Ann winning by the first way (namely, by H coming up on toss 1) are

HHH, HHT, HTH, HTT

so indeed the probability is 4/8, or 1/2, for her winning by the first way. Also the chances favorable to Ann winning by the second way (namely, by H coming up at least 2 times out of 3 tosses) are

HHH, HHT, HTH, THH

so the probability is 4/8, or 1/2, for her winning by the second way. For 3 of these chances, namely

HHH, HHT, HTH,

we see that Ann wins directly by the first way (namely, H on toss 1) and so toss 2 and toss 3 need not be made. Thus the two ways that Ann might win are not incompatible, so it was not legitimate to add their probabilities. In other words) the chances

HHH, HHT, HTH

are common to both ways. The chances favorable to Ann winning one way or the other are

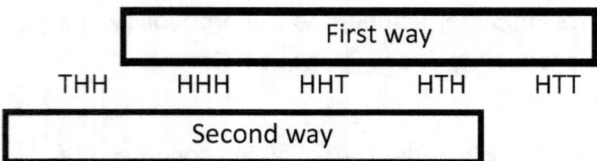

Thus her chances are 5 in 8. Thus the probability of Ann winning is 5/8.

CHAPTER 13. MULTIPLICATION RULE

Blaise Pascal wrote: No one passes in the world as skilled in verse unless he has put up the sign of a poet, a mathematician, &c. But educated people do not want a sign, and draw little distinction between the trade of a poet and that of an embroiderer.

13.1 The multiplication rule

Another general rule, which like the addition rule is extremely useful, is the multiplication rule. It is

RULE. If one thing can be done in M different ways, and then another thing can be done in N different ways, it follows that both things can be done in the product of M × N different ways.

Another statement of the multiplication rule is

RULE. If there is an M-way choice for thing 1 and then there is an N-way choice for thing 2, it follows that there is an MN-way choice for both things.

This rule is basic. It was first taught to you when you learned how to multiply numbers. For example, if you first choose one of coffee, tea, or milk, and then either a ham sandwich or a peanut butter sandwich, there are 3 × 2 = 6 different choices that you have, namely

coffee, ham sandwich	coffee, peanut butter sandwich
tea, ham sandwich	tea, peanut butter sandwich
milk, ham sandwich	milk, peanut butter sandwich

Suppose there are 3 trails to the top of Mount Washington. Thus there are 3 ways up and 3 ways down, so that there are 3 × 3, or 9 ways of making one round trip. If the 3 trails are called A, B, C, then these 9 round trips are

AA	AB	AC
BA	BB	BC
CA	CB	CC

Suppose now that we do not want to come down by the same trail as the trail we went up. Now there are 3 trails to go up, but only 2 trails that we can come down on. Hence the number of round trips is 3×2, or 6. They are

AB	AC
BA	BC
CA	CB

These 6 round trips could be obtained by striking out the 3 ineligible ones, namely AA, BB, CC in the list of 9 above.

13.2 Some problems on multiplication and solutions.

Problem 1. A friend shows me 4 detective books and 8 western books. I can choose one of each. What choices have I?

Solution. 4×8 = 32

Problem 2. In how many ways can we select a consonant and a vowel out of the English alphabet?

Solution. There are 20 consonants and 6 vowels, so there are 20×6, or 120 ways.

Problem 3. In how many ways can we form a 2-lettered word out of 26 letters, the 2 letters of the word being different?

Solution. Letter 1 can be chosen in 26 ways, and thereafter letter 2 can be chosen in 25 ways. Thus there are 26×25, or 650 such words.

Problem 4. Two persons get into an airplane where there are 9 vacant seats. In how many different ways can they seat themselves?

Solution. Person 1 can take any of the 9 vacant seats. Thereafter, person 2 can take any of the 8 seats that are left. Hence there are 9×8, or 72 different ways that they can take their seats.

13.3 Some problems on probability and solutions.

Blaise Pascal wrote: The strength of a man's virtue must not be measured by his efforts, but by his ordinary life.

The multiplication rule is very useful in computing probabilities. In the problems here we make use of the fundamental formula:

Probability = (Number of favorable chances)/ (Number of all chances)

Problem 5. Two cards are drawn from a well-shuffled pack of 52 cards. What is the probability that both the cards drawn are kings?

Solution. Since there are 52 cards in the pack, the first card can be drawn in 52 ways. After the first card has been withdrawn, there are 51 cards remaining in the pack, so the second card can be drawn in 51 ways. Therefore the total number of ways to draw 2 cards is 52×51 = 2652, which is the number of all the chances. To find the number of chances favorable to the event of drawing 2 kings, we observe that there are 4 kings in the pack. Thus the first king can be drawn in 4 ways. After the first king has been drawn, there are 3 kings left, so the second king can be drawn in 3 ways. Therefore the number of favorable chances is 4×3 = 12. Hence there are 12 chances in 2652 that both cards drawn are kings, so the required probability is

P(2K) = 12/2652 = 1/221

That is, there is 1 chance in 221 of drawing 2 kings if the first card is not returned to the pack before drawing the second card.

Problem 6. Two cards are drawn from a pack of 52 cards, the first card being returned to the pack before the second card is drawn. (By this description we mean: The cards are shuffled and the first card is drawn and noted. This card is then returned to the pack, the pack is shuffled again, and the second card is drawn and noted.) What is the probability that both cards drawn are kings?

Solution. There are 52 ways of drawing the first card. There are also 52 ways of drawing the second card, because by returning the first card drawn, the pack is restored to its original number of 52 cards. Thus the number of ways to draw both cards is 52×52, which is 2704. There are 4 ways of drawing the first king, and because the pack is restored, there are also 4 ways of drawing the second king. Thus the number of ways to draw both kings is 4×4, which is 16. Hence the required probability is

P(2K) = 16/2704= 1/169

In other words,, there is 1 chance in 169 of drawing 2 kings if the first card drawn is returned to the pack before drawing the second card.

Problem 7. There are 2 red balls (labeled R1, R2) and 3 black balls (labeled B1, B2, B3) in a bag. One ball is drawn. What is the probability that it is black?

Solution. There are 5 balls in total, which represent all the chances. Since there are 3 chances favorable to a black ball being drawn, the required probability is

P(B) = 3/5 .

Problem 8. The contents of the bag are the same as the foregoing problem; namely, R1, R2, B1, B2, B3. For this problem, however, we suppose that one ball is drawn, its color is unnoted, and it is laid aside. Then another ball is drawn. What is the probability that the second ball drawn is B given that the first ball drawn is unknown? (Note. We denote this probability by P(B given ?).

Solution 1. There are 5 ways to draw the first ball, and, whatever it is, then are only 4 ways to draw the second ball. Thus there are 5×4, or 20 chances in all; namely,

R1 R2	R2 R1	B1 R1	B2 R1	B3 R1
(R1 B1)	(R2 B1)	B1 R2	B2 R2	B3 R2
(R1 B2)	(R2 B2)	(B1 B2)	(B2 B1)	(B3 B1)
(R1 B3)	(R2 B3)	(B1 B3)	(B2 B3)	(B3 B2)

The 12 favorable chances are in parentheses. The required probability is

P(B given ?) = 12/20 = 3/5.

Solution 2. There are 2 incompatible ways in which the second ball drawn may be black; namely, way 1 and way 2.

Way 1. Suppose that the first ball drawn is black, so it must be one of B1, B2, B3. Whichever one this black is, the second ball drawn, if it is black, must be one of the 2 remaining black balls. Therefore, under the condition that the first is black, there are 3×2 = 6 favorable chances. (Note. From the foregoing Solution 1, these favorable chances are

(B1 B2)	(B2 B1)	(B3 B1)
(R1 B3)	(B2 B3)	(B3 B2)

Way 2. We suppose that the first ball drawn is red, so it must be one of R1, R2. Whichever one this red ball is, the second ball drawn, if it is

black, must be one of the 3 black balls, namely Bl, B2, B3. Therefore under the condition that the first ball drawn is red, there are 2×3 = 6 favorable chances. (Note. From the foregoing Solution 1, these favorable chances are

(R1 B1)	(R2 BL)
(R1 B2)	(R2 B2)
(R1 B3)	(R2 B3)

Because Way 1 and Way 2 are incompatible the addition rule tells us that the number of favorable chances is 6 + 6 = 12. Since there are 5×4 = 20 chances in all, the required probability is therefore P(B given ?) = 12/20 = 3/5. (Note. It is remarkable that the probability of black on the first draw has the same numerical value as the probability of black on the second draw, given that the outcome of the first draw is unknown. In other words, it is remarkable that

$$P(B) = P(B \text{ given } ?) = 3/5.$$

However, the situation is different if we know the outcome of the first draw, as illustrated in the following 2 problems.

Problem 9. The contents of the bag are the same as the foregoing problem, namely R1, R2, B1, B2, B3. For this problem, however, we suppose that one ball is drawn, its color is noted, it is red and it is laid aside. Then another ball is drawn. What is the probability that the second ball drawn is black, given that the first ball drawn is red?

Note. We denote this probability by P(B given R), which is read "P of B given R."

Solution. Since there are 2 red balls in the bag, namely R1, R2, there are 2 ways that the first (red) ball can be drawn. Since there are then 4 balls left, there are 4 ways that the second ball can be drawn. Thus the number of all chances is 2×4 = 8. As we have just seen, there are 2 ways that the first red ball can be drawn. Since there are then 3 black balls left, namely B1, B2, B3, there are 3 ways that the second ball drawn can be black. Thus the number of favorable chances is 2×3 = 6. The required probability is therefore

P(B given R) = 6/8 = 3/4

Problem 10. The contents of the bag are the same as the foregoing problem; namely R1, R2, B1, B2, B3. For this problem, however, we suppose that one ball is drawn, its color noted, it is black, and it is laid aside. Then another ball is drawn. What is the probability that the second ball drawn is black, given that the first ball drawn is black?

Note. We denote this probability by P(B given B) which is read "P of B given B."

Solution. The number of all chance is 3×4 = 12. The number of favorable chances is 3×2 = 6. The required probability is therefore

P(B given B) = 6/12 = 1/2.

This problem, and the foregoing problem, illustrate that probability depends upon the given set of conditions.

Problem 11. 3 chests, identical in appearance, each have 2 drawers. Chest 1 has a G (gold) coin in each of drawers 1 and 2. Chest 2 has a S (silver) coin in each of drawers 1 and 2. Chest 3 has a S (silver) coin in drawer 1 and a G (gold) coin in drawer 2. See Figure 1.

	Chest 1	Chest 2	Chest 3
Drawer 1	G	S	S
Drawer 2	G	S	G

Figure 1. The three chests.

A chest is chosen haphazardly. What is the probability that its 2 coins are of different metals?

Solution. Since outwardly the chest are indistinguishable from each other, we recognize each chest as having 1 chance to be drawn. Among the 3 chests, only 1 chest has coins of different metals, namely chest 3. Therefore the required probability is 1/3.

Problem 12. Let there be 3 chests as in the foregoing problem. A chest is chosen haphazardly, one of its drawers is opened, and a S (silver) coin is found. What is the probability that the other drawer contains a G (gold) coin?

Solution. The fact that S was found in one drawer leaves only 2 possibilities as to the content of the other drawer; namely S, G. Hence we might be tempted to reason that the probability of G in the other drawer is 1/2. Nevertheless, this reasoning is false, because each of the possibilities cannot be regarded as having 1 chance each. Before the chest is chosen and one drawer opened, there are 6 possible results.

Drawer 1 of chest 1	Drawer 1 of chest 2	Drawer 1 of chest 3
Drawer 2 of chest 1	Drawer 2 of chest 2	Drawer 2 of chest 3

Each one of the 6 represents 1 chance. However, as soon as S is found in the drawer that is opened, the 3 cases

Drawer 1 of chest 1		
Drawer 2 of chest 1		Drawer 2 of chest 3

become impossible, and so there remain only 3 cases, namely

	Drawer 1 of chest 2	Drawer 1 of chest 3
	Drawer 2 of chest 2	

Each of these cases represents 1 chance. In these 3 chances, there is 1 chance, namely Drawer 1 of chest 3, that the other drawer contains G. There the required probability is 1/3.

13.4 Tree diagrams.

The multiplication rule can be displayed graphically by means of a tree diagram.

Problem 13. A man has 2 suits (gray and brown) and 3 neckties (red, blue and green). What choices of suit and necktie does he have?

Solution. His choice is 2×3, or 6. To illustrate the use of a tree diagram, we let the left fork represent the choice of a suit. There are 2 suits so the left fork has 2 branches. Thus this fork is a 2-way fork. Each of the 2 branches on the left fork leads to a separate right fork. Each of the 2 right forks represents the choice of a necktie. There are 3 neckties se each right fork has 3 branches. Thus each right fork is a 3-way fork. A choice of both suit and necktie correspond to a choice of path from the left to right.

As we see in Figure 1, there are 2×3 = 6 such paths.

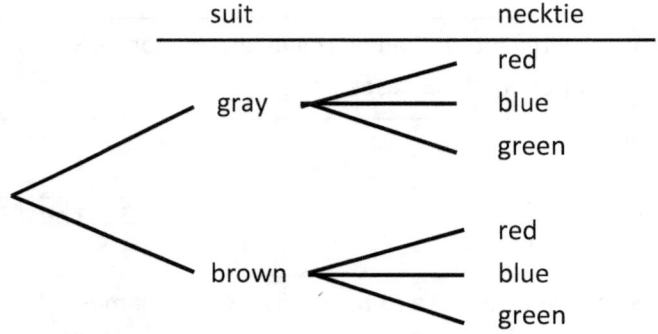

Figure 1. Tree diagram with 6 paths.

Let us now consider the tree diagram for the multiplication rule for 3 things.

Suppose thing 1 can be done in N1 different ways, then thing 2 can be done in N2 different ways),and then thing 3 can be done in N3 different ways.

The fork to thing 1 is an N_1 -way fork.

For each branch of this N_1 -way fork there is a fork to thing 2.

Each of these forks to thing 2 is an N_2-way fork.

For each branch of these N_2-way fork there is a fork to thing 3.

Each of those forks to thing 3 is an N_3-way fork.

The number of paths of the form "thing 1 then thing 2 then thing 3" is equal to $N_1 N_2 N_3$.

For example, the tree diagram for $N_1 = 3$, $N_2 = 2$, $N_3 = 4$ is shown in Figure 2.

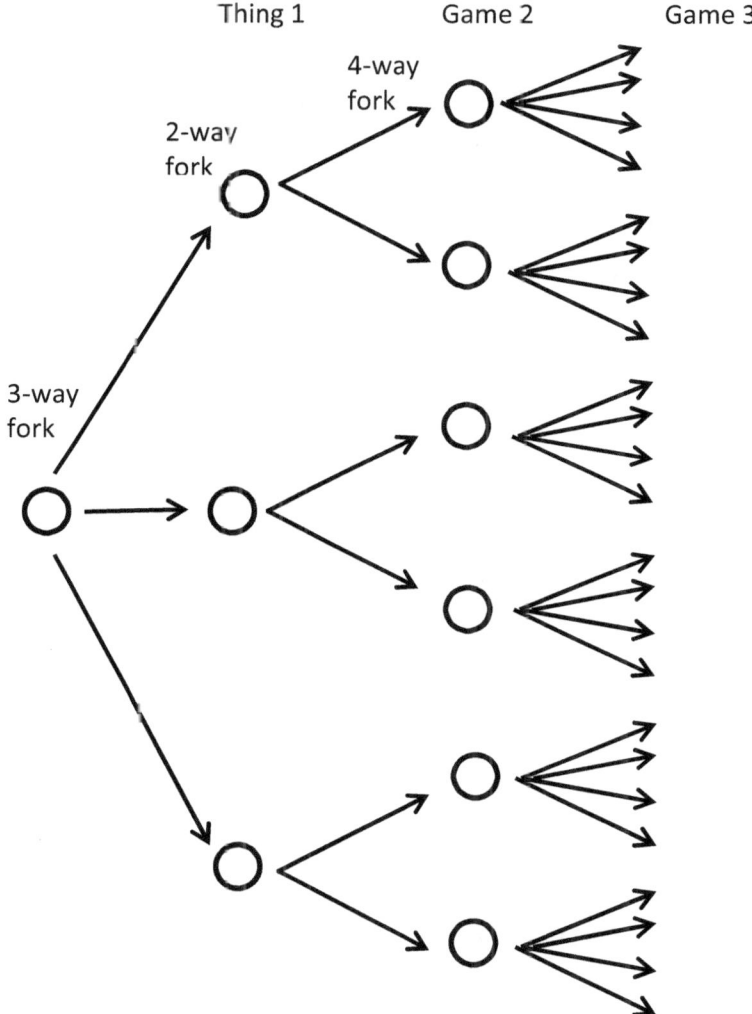

Figure 2. Tree diagram with 3×2×4=24 paths

The multiplication rule may be stated in terms of a tree diagram as follows.

A fork is a set of branches. The fork to thing 1 has N_1 branches, any fork to thing 2 has N_2 branches, ... , any fork to thing k has N_k branches. To each branch of the fork to thing 1 is attached a fork to thing 2, to each branch of the fork to thing 2 is attached a fork to thing 3, ... , to each branch of the fork to thing k −1 is attached a fork to thing k. This entire set of forks makes up the tree. A path is a series of branches. A path is made up of one branch from the fork to thing 1, one branch from the corresponding fork to thing 2, . . . , one branch from the corresponding fork to thing k. Then the multiplication rule says that the number of different paths is equal to the product $N_1 \ N_2 \cdots N_k$ .13.4 General multiplication rule.

We may extend the multiplication rule to doing 3 things, one after another. Thus, we have:

RULE. If thing 1 can be done in N_1 different ways, and thereafter thing 2 can be done in N_2 different ways, and thereafter thing 3 can be done in N_3 different ways, it follows that all the things can be done in $N_1 \ N_2 \ N_3$ different ways.

Problem 14. 10 persons compete for 3 prizes. In how many different ways can the prizes be awarded?

Solution. Prize 1 can be given in 10 different ways. Thereafter prize 2 can be given in 9 different ways. Thereafter prize 3 can be given in 8 different ways. Hence there are (10) (9) (8) = 720 different ways to give the prizes.

Problem 15. In how many ways can 3 letters be put into 3 addressed envelopes, one letter into each envelope.

Solution. There are 3 ways of filling envelope 1. Then there are 2 letters left, so there are 2 ways of filling envelope 2. Now there is only 1 letter left so it must go into envelope 3, and hence there is only way to fill envelope 3. Thus there are

(3) (2) (1) = 6 ways of doing the whole job.

The multiplication rule can be extended to doing any number of things.

RULE. Suppose a series of things can be done successively as follows.
Thing 1 in N_1 different ways,
Thereafter thing 2 in N_2 different ways,
Thereafter thing 3 in N_3 different ways, etc.,
Thereafter then thing k in N_k different ways
It follows that all the things can be done in
$N_1 \times N_2 \times N_3 \times \ldots \times N_k$ different ways.

Another statement of the multiplication rule is:

> **RULE**. Suppose a chain is made up of k successive links, where link 1 can be ore of N_1 different kinds, link 2 can be one of N_2 different kinds, link 3 can be one of N_3 different kinds, etc., until link k can be one of N_k different kinds. The conclusion is that there are $N_1 N_2 N_3 \cdots N_k$ different kinds of chains.

Still another statement of the multiplication rule is:

> **RULE**. If a path is made up of k successive things, where
>
> thing 1 has can be one of N_1 different branches,
>
> thing 2 has can be one of N_2 different branches,
>
> thing 3 has can be one of N_3 different branches, etc.,
>
> thing k has can be one of N_k different branches,
>
> then there are $N_1 N_2 N_3 \cdots N_k$ different paths.

Blaise Pascal wrote: Eloquence is an art of saying things in such a way—(1) that those to whom we speak may listen to them without pain and with pleasure; (2) that they feel themselves interested, so that self-love leads them more willingly to reflection upon it.

It consists, then, in a correspondence which we seek to establish between the heac and the heart of those to whom we speak on the one hand, and, on the other, between the thoughts and the expressions which we employ. This assumes that we have studied well the heart of man so as to know all its powers, and then to find the just proportions of the discourse which we wish to adapt to them. We must put ourselves in the place of those who are to hear us, and make trial on our own heart of the turn which we give to our discourse in order to see whether one is made for the other, and whether we can assure ourselves that the hearer will be, as it were, forced to surrender. We ought to restrict ourselves, so far as possible, to the simple and natural, and not to magnify that which is little, or belittle that which is great. It is not enough that a thing be beautiful; it must be suitable to the subject, and there must be in it nothing of excess or defect.

CHAPTER 14. STOCHASTIC PROCESSES

Blaise Pascal wrote: We know the truth not only through our reason but also through our heart. It is through the latter that we know first principles, and reason, which has nothing to do with it, tries in vain to refute them.

14.1 Stochastic processes.

Two or more stochastic phenomena in a row make up a stochastic process. For example, each toss of a coin represents a stochastic phenomenon. Hence repeated tosses of the coin would make up a stochastic process. The stochastic process made up of 3 tosses has the possible results, or cases, shown in the Figure.

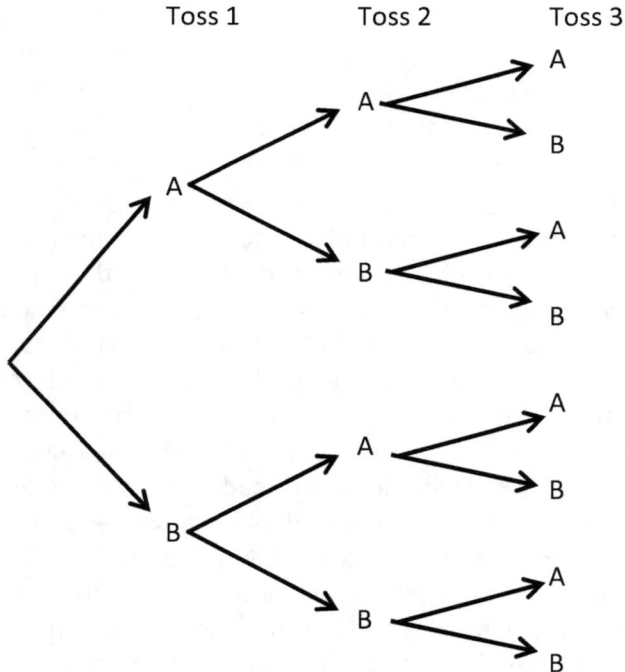

Figure 1. Stochastic process made up of 3 tosses of a coin.

Each path in the Figure is a possible result or case of the stochastic process. There are 8 such paths. For example, one path is H T H. A stochastic process considered as a whole is itself a stochastic phenomenon. Thus the above example of 3 repeated tosses of a coin is a stochastic phenomenon with the possible results, or cases, given by

HHH, HHT, HTH, HTT, THH, THT, TTH, TTT

Hence we always have the option to treat a stochastic process either

(1) as 2 or more individual phenomena in a row, or

(2) as one overall phenomenon.

14.2 Some problems and solutions.

Problem 1. Purse 1 contains 1 dime and 2 nickels. Purse 2 contains 2 dimes and 1 nickel. You happen to take a purse and a coin from it. That is the probability that the coin is a dime.

Solution.

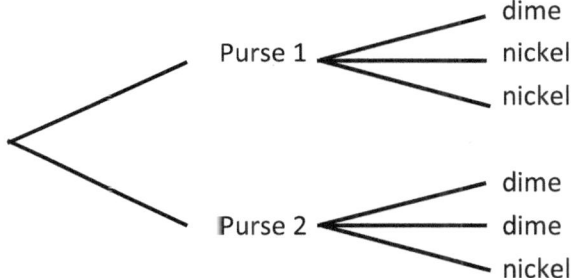

Figure 2. Stochastic process made up of taking a purse and a coin from the chosen purse.

The first stochastic phenomenon is the taking of a purse; the second stochastic phenomenon is the taking of a coin from it. Both of these phenomena make up the stochastic process illustrated in the Figure. From the Figure we see there are 6 paths, each of which may be considered to represent 1 chance. Of these 6 chances, 3 result in a dime. Hence there are 3 chances in 6, or a probability equal to 3/6 = 1/2, that the coin is a dime.

Exercising our other option, we may treat the situation as one overall stochastic phenomenon. This overall phenomenon has 6 possible results; namely,

purse 1	then dime
purse 1	then nickel
purse 1	then nickel
purse 2	then dime
purse 2	then dime
purse 2	then nickel

Each of the above six items may be considered as 1 chance. The probability of a dime is, as before, 3/6 = 1/2.

Alternative solution. Instead of constructing a tree diagram so that each path represents 1 chance, we can construct a tree diagram as shown in Figure 3.

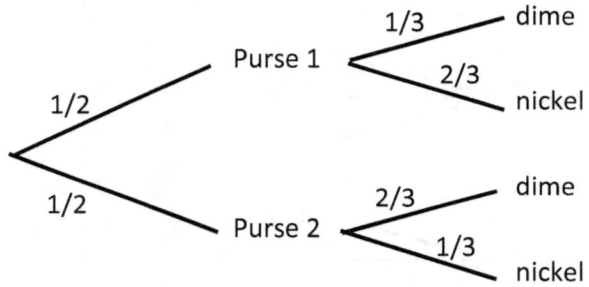

Figure 3: Stochastic process made up of taking a purse and a coin from the purse taken.

In this tree diagram, we have assigned a probability to each branch. For the first phenomenon, each purse has 1 chance in 2 to be taken, so we have assigned a probability of 1/2 to each branch.

For the second phenomenon as must consider 2 situations, namely either purse 1 or purse 2 was taken. Given that purse 1 was taken, there is 1 chance in 3 for a dime, so the probability 1/3 is assigned to the "dime" branch leading out of purse 1. Given that purse 1 was taken, there are 2 chances in 3 for a nickel, so the probability 2/3 is assigned to the "nickel" branch leading out of Purse 1.

Given that purse 2 was taken, there are 2 chances in 3 for a dime, so the probability 2/3 is assigned to the "dime" branch leading out of purse 2. Given that purse 2 was taken, there is 1 chance in 3 for a nickel, so the probability 1/3 is assigned to the "nickel" branch leading out of purse 2.

Let us consider the sequence of the 2 events:

Event (1)   purse 1 is taken

Event (2)   a dime is taken from it.

This sequence of 2 events, which is the top path in the Figure, may be denoted by purse 1 then dime.

The probabilities on the branches of this path are:

Probability = 1/2   on the branch purse 1,

Probability = 1/3   on the branch dime, given purse 1.

We see that the probability of the path corresponding to "purse 1" on the first phenomenon and "then dime" on the second phenomenon is equal to the product of the probabilities associated with each branch along the path. We have:

Probability of "purse 1 then dime" is (1/2)(1/3) = 1/6

Probability of "purse 1 then nickel" is (1/2)(2/3) = 2/6

Probability of "purse 2 then dime" is (1/2)(2/3) = 2/6

Probability of "purse 2 then nickel" is (1/2)(1/3) = 1/6

Exercising our option to consider the stochastic process as an overall stochastic phenomenon, we have the 4 possible results. Each of these four possible results makes up a simple event with probabilities

P(purse 1 then dime)   (1/2)(1/3) = 1/6

P(purse 1 then nickel)   (1/2)(2/3) = 2/6

P(purse 2 then dime)   (1/2)(2/3) = 2/6

P(purse 2 then nickel)   (1/2)(1/3) = 1/6.

The event of taking a dime is the union of the two events

Event 1   purse 1 then dime,

Event 2   purse 2 then dime,

which are incompatible. That is,

dime = (purse 1 then dime) or (purse 2 then dime)

where

(purse 1 then dime) and (purse 2 then dime) = None.

Hence by the theorem of total probability, we have

P(dime)=P(purse 1 then dime)+P(purse 2 then dime)

=(1/6)+(2/6)=1/2.

Problem 2. Purse 1 contains 1 dime and 1 nickel. Purse 2 contains 2 dimes and 1 nickel. You happen to take a purse and a coin from it. What is the probability that the coin is a dime?

Solution. Figure 3 shows the tree diagram with branch probabilities.

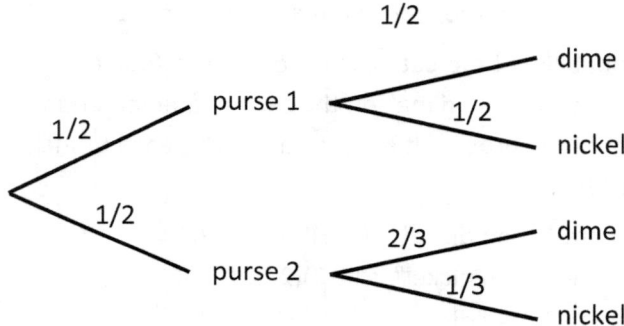

Figure 4. Tree diagram with branch weights.

The desired probability is

P(dime)

= P(purse 1 then dime) + P(purse 2 then dime)

= (1/2)(1/2)+(1/2)(2/3) = (1/4) + (1/3)

= (3/12) + (4/12) = 7/12.

Alternate solution. Let us construct a tree diagram so that each path is 1 chance. Since purse 1 and purse 2 are equally probable, there must be the same number of paths going through each of purse 1 and purse 2.

To accomplish this end, we suppose that purse 1 contains 3 dimes and 3 nickels, and purse 2 contains 4 dimes and 2 nickels. (It is seen that this supposition does not change the branch probabilities shown on the tree diagram in the Figure 4.)

Under this supposition, we have the tree diagram shown in Figure 5.

Each path is this tree diagram represents 1 chance. Since 7 paths result in a dime out of the 12 paths in all, we see that P(dime) = 7/12, which is the same answer that we obtained in the foregoing solution.

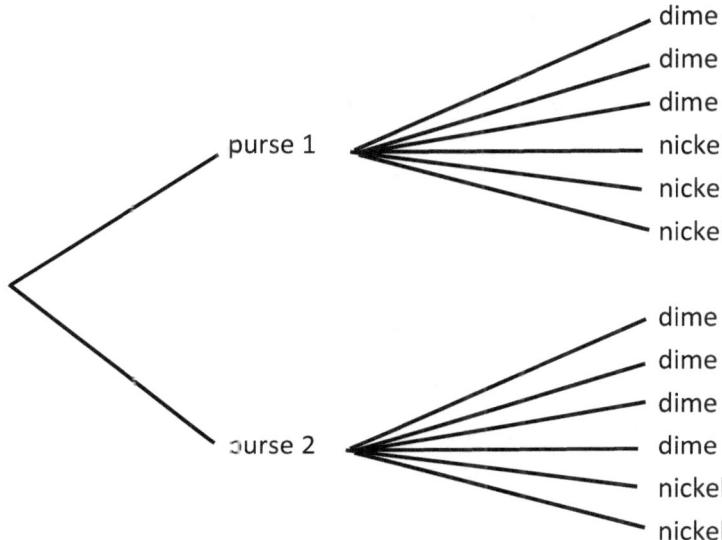

Figure 5: Tree diagram where each path is 1 chance.

Problem 3. Suppose there are 2 urns. Urn 1 contains 2 apples and 1 orange. Urn 2 contains 2 apples and 3 oranges. You happen to take an urn and a fruit from it. What is the probability that the fruit is an apple?

Solution. The tree diagram with branch probabilities is shown in Figure 6.

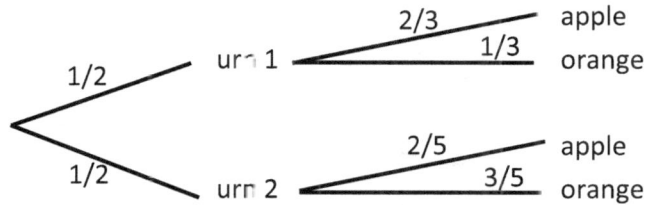

Figure 6. Tree diagram with branch probabilities.

The probability that an apple is taken is equal to the sum of the probabilities of the paths that lead to an apple. That is,

P(apple) P(urn 1 then apple) + P(urn 2 then apple)

= (1/3)(2/3) + (1/2)(2/5) = 1/3 + 1/5 = 8/15.

Problem 4. The probability that a newspaper reader will read an advertisement is 0.3 and if he reads the advertisement the probability that he will buy the product advertised is 0.02. What is the probability that he will read the advertisement and then buy the product?

Solution. The tree diagram is shown in Figure 7.

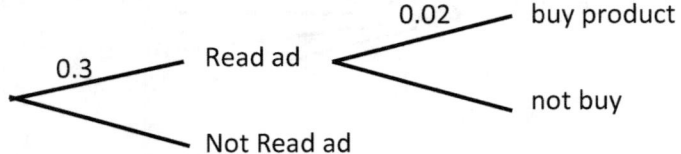

Figure 7. Tree diagram with branch probabilities.

The required probability is (0.3)(0.02) = 0.006.

Problem 5. If a letter is drawn from (and not returned to) an alphabet of 20 consonants and 6 vowels and the letter drawn is a vowel, what is the probability that another letter drawn is a vowel.

Solution. After the first drawing, there are 25 letters left. Since it is given that the first drawing was a vowel, it follows that 5 of these 25 letters are vowels, and the other 20 are consonants. Hence the probability that another letter drawn is a vowel (given that the first letter drawn was a vowel) is

P(letter 2 is a vowel given letter 1 is a vowel) = 5/25 = 1/5

Problem 6. What is the probability of drawing 2 vowels from an alphabet of 20 consonants and 6 vowels?

Solution. The tree diagram is shown in Figure 8.

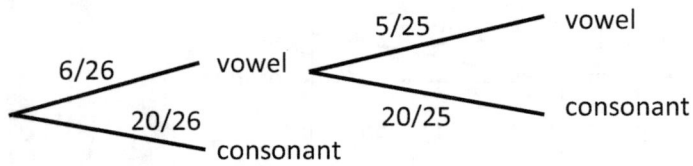

Figure 8. Tree diagram with branch probabilities.

Hence the required probability is

P(letter 1 is a vowel then letter 2 is a vowel)= (6/26)(5/25)=3/65

We may verify this answer by observing that a drawing of 2 letters can be made in 26×25 ways, and a drawing of 2 vowels can be made in 6×5 ways, so the required probability is (6×5)/(26×25) = 3/65, as found above.

Problem 7: A coin is tossed. If it falls H, then it is tossed again. If it falls T, then a die is tossed. What is the probability of an ace on the die?

Solution. The tree diagram is shown in Figure 9.

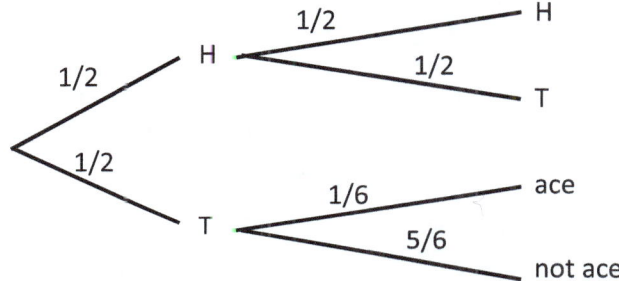

Figure 9.Tree diagram with branch probabilities.

The required probability is

   (1/2)(1/6) = 1/12

Problem 7. Purse 1 contains 4 nickels and 3 dimes. Purse 2 contains 4 nickels and 5 dimes. You happen to take a purse and a coin from it. What is the probability that it is a nickel?

Solution. The probability that purse 1 is picked is 1/2, and if it is picked the probability that the coin is a nickel is 3/7. Therefore the probability that the coin drawn is one of the nickels in purse 1 is

   (1/2) (4/7) = 2/7

Similarly, the probability that the coin drawn is one of the nickels in purse 2 is

   (1/2)(4/9) = 2/9

Hence the probability of drawing a nickel is

   2/7 + 2/9 = 18/63 + 14/63 = 32/63.

Problem 8. What is the expectation from the drawing of the foregoing problem?

Solution. A nickel is $0.05. The probability that the coin drawn is a nickel is 32/63, so its expectation is

(32/63) ($0.05) = $16/630

A dime is $0.10. If the coin drawn is not a nickel, then it must be a dime. Thus the probability of drawing a dime is 1 - 32/63 = 31/63, so its expectation is

(31/63) ($0.10) = $31/630

Thus the expectation from the drawing is

16/630 + 31/630 = 47/630 = $0.0746

Problem 9. Suppose that the coins of the foregoing 2 problems are put into 1 purse. That purse will contain 8 nickels and 8 dimes. If you happen to draw a coin, what is your expectation?

Solution. The probability of drawing a nickel is 8/16=1/2, which is a little less than before, and the probability of drawing a dime is 8/16 = 1/2, which is a little more than before. The expectation is

(1/2)×($0.05)+(1/2)×($0.10)=$0.075

which is a little more than before,

Problem 10. Before moving to another city, a person stored the books that he did not want to take with him in 3 sealed boxes and left them behind. Now he wants a particular book that is in one of the 3 boxes. The probability that it is in one particular box is 3/5. But if it is not in that box, it is equally probable to be in either of the other 2. If he sends for this particular box, giving a description of it, the probability of getting the box he describes is 2/3. What is his probability of getting the book he wants?

Solution. The probability of getting the box described is 2/3, and the probability that the desired book in it is 3/5. Thus the probability of getting the book in this way is (2/3)(3/5) = 6/15.

The probability of getting a box not described is 1/3, and the probability that the book is in it is 1/5. Thus the probability of getting the book in this way is (1/3)(1/5) = 1/15.

Therefore the probability of getting the book one way or the other is

6/15 + 1/15 = 7/15.

Alternative solution. The tree diagram is shown in Figure 10.

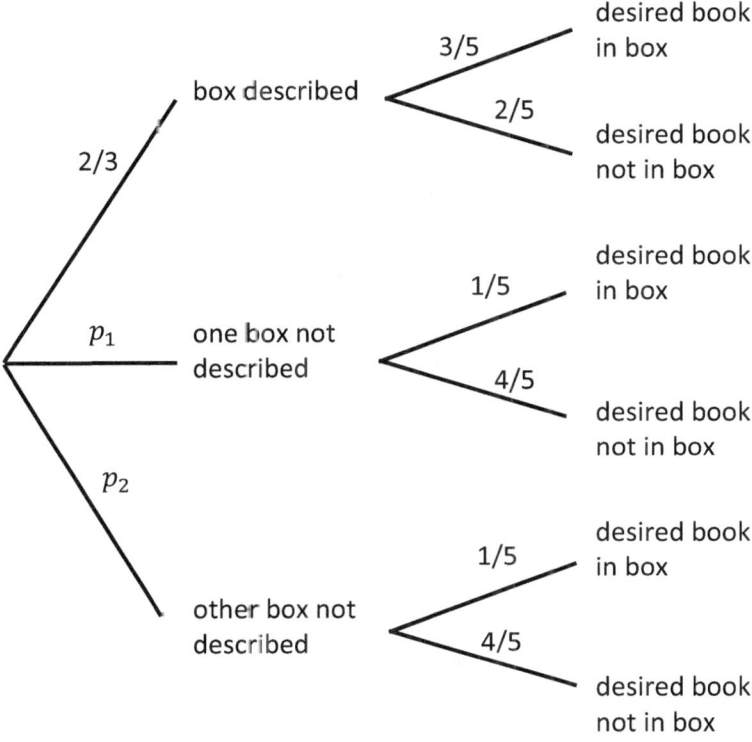

Figure 10. Tree diagram with branch probabilities.

The branch probabilities p_1, p_2 are not known, but we do know that

   $p_1 + p_2 = 1/3$.

The required probability is

   $(2/3)(3/5) + p_1 (1/5) + p_2 (1/5)$

   $= (2/3)(3/5) + (p_1+p_2)(1/5)$

   $= (2/5)(3/5) + (1/3)(1/5)$

   $= 6/15 + 1/15 = 7/15$

which is the same answer as in the foregoing solution.

14.3 More Problems

Problem 11. Purse 1 contains a nickel and a dime. Purse 2 contains 2 nickels. You happen to draw a coin from purse 1 and put it into purse 2.

Then you happen to draw a coin from purse 2 and put it into purse 1. You now ask a friend to pick whichever purse he pleases. Which purse should he pick?

Solutions. Lot N stand for nickel ($0.05) and lot D stand for dime ($0.10). The tree diagram is shown in Figure 11.

Start: purse 1 N, D; purse 2, N, N	Coin that leaves purse 1	Coin that returns to purse 1	End result in purse 1

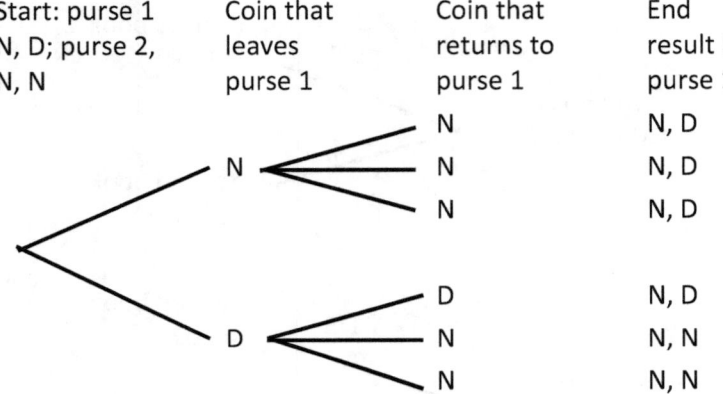

	N	N, D
N	N	N, D
	N	N, D
	D	N, D
D	N	N, N
	N	N, N

Figure 11. Tree diagram for which each path is 1 chance.

The only way for D to be in purse 2 is that the coin that leaves purse 1 is D and that the coin that returns to purse 1 is not D. We see that there are 2 such paths out of 6, so the probability that D is in purse 2 at the end is

P(D in purse 2 at the end)= 2/6 = 1/3.

Likewise,

P(D in purse 1 at the end)= 1 – (2/6) = 4/6 = 2/3.

Thus your friend should pick purse 1 and not purse 2.

Problem 12. 2 dice are tossed. What is the

(a) probability that they should fall alike

(b) probability that they should fall different.

Solution. The tree diagram is shown in Figure 12.

The probability that they should fall alike is

6 × (1/6) (1/6) = 6×(1/36) = 1/6 .

The probability that they should fall different is

6×(1/6) (5/6) = 6×(5/36) = 5/6.

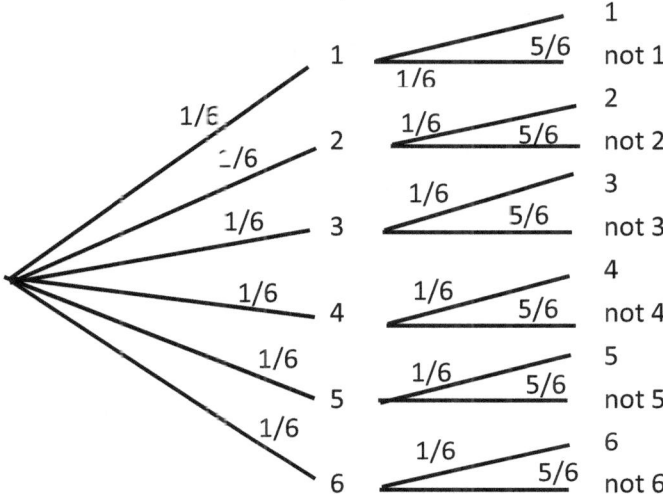

Figure 12. Tree diagram with branch probabilities.

Problem 13. 3 dice are tossed. Find

(a)   probability that the 3 should all fall alike

(b)   probability that only 2 should fail alike

(c)   probability that the 3 should all fall different

Solution.

(a) The probability that die 2 should fall the same as die 1 is 1/6. The probability that die 3 should also fall the same is 1/6. Therefore the probability that the 3 dice should all fall alike is (1/6)(1/6) = 1/36.

(b) The probability that die 1 and die 2 should fall alike and that die 3 should fall d fferent s

(1/6)(5/6) = 5/36.

The probabi ity 5/36 also is the probability that die 2 and die 3 should fall alike, and that die 1 different. The probability 5/36 also is the probability that die 1 and die 3 should fall alike, and that die 2 different.

Therefore the probability that 2 dice should be alike and the other different is

5/36 + 5/36 + 5/36 = 15/36

(c) The probability that die 2 should be different from die 1 is 5/6, and the probability that die 3 should be different from either die 1 or die 2 is 4/6. Therefore the probability that the 3 dice are different is

(5/6)(4/6) = 20/36

As a check, we see that the 3 desired probabilities add up to 1, that is,

1/36 + 15/36 + 20/36 = 1

That they add up to 1 is necessary because the 3 dice have to fall in one of the 3 ways.

Problem 14. A person throws 3 dice. He receives $6 if they all fill alike, $4 if only 2 fall alike, and $3 if they all fall different. What is his expectation?

Solution. Using foregoing problem, we see that his expectation is

(1/36)($2) + (15/36)($4) + (20/36)($3) = $3.50

Problem 15. A bag contains 5R (red) balls and 3B (black) balls. You happen to draw 2 balls from it, one after the other, without replacing ball 1 before ball 2 is drawn. Find

(a)   probability that ball 1 is R and ball 2 is B
(b)   probability that ball 1 is B and ball 2 is R
(c)   probability that one ball is R and the other B
(d)   probability that both bells are R
(e)   probability that both balls are B

Solution. The tree diagram is shown in Figure 13.

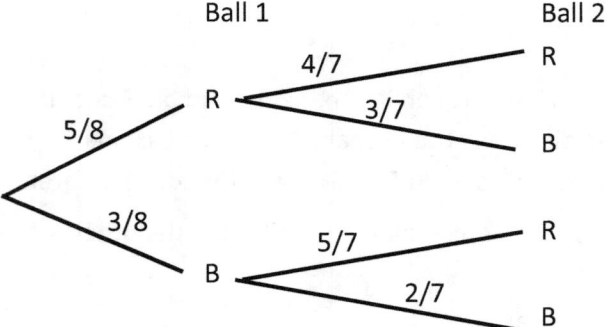

Figure 13 Tree diagram with branch probabilities.

The required probabilities are:

(a) $(5/8)(3/7) = 15/56$
(b) $(3/8)(5/7) = 15/56$
(c) $(15/56)(15/56) = 30/56$
(d) $(5/8)(4/7) = 20/56$
(e) $(3/8)(2/7) = 6/56$

Check: $15/56 + 15/56 + 20/56 + 6/56 = 56/56 = 1$.

We note that the drawing of ball 2 is dependent on the outcome of the drawing of ball 1.

This dependence is shown by Figure 13. We see that the fork leading out of R for ball 1 is different than the fork leading out of B for ball 1.

Problem 16. A bag contains 5R (red) balls and 3B (black) balls. You happen to draw 2 balls from it, one after the other, replacing ball 1 before ball 2 is drawn. Find

(a)   the probability that ball 1 is R and ball 2 is B
(b)   the probability that ball 1 is B and ball 2 is R
(c)   the probability that one ball is R and the other B
(d)   the probability that both balls are R
(e)   the probability that both balls are B

Solution. The tree diagram is shown in Figure 14.

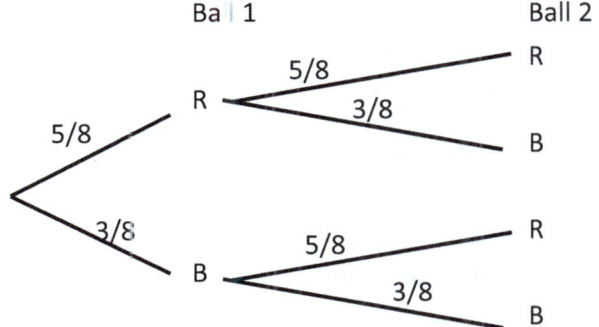

Figure 14. Tree diagram with branch probabilities.

The required probabilities are

(a) $(5/8)(3/8) = 15/64$

(b) $(3/8)(5/8) = 15/64$

(c) $(15/64) + (15/64) = 30/64$

(d) $(5/8)(5/8) = 25/64$

(e) $(3/8)(3/8) = 9/64$

Check: $15/64 + 15/64 + 25/64 + 9/64 = 64/64 = 1$.

We note that the drawing of ball 1 is independent of the drawing of ball 2. In other words, the 2 drawings have no influence on each other, as seen by the fact that both forks leading out of ball 1 in the Figure are the same.

Problem 17. 2 dice are tossed. What is the probability that only 1 die shows an ace?

Solution. The tree diagram is shown in Figure 15.

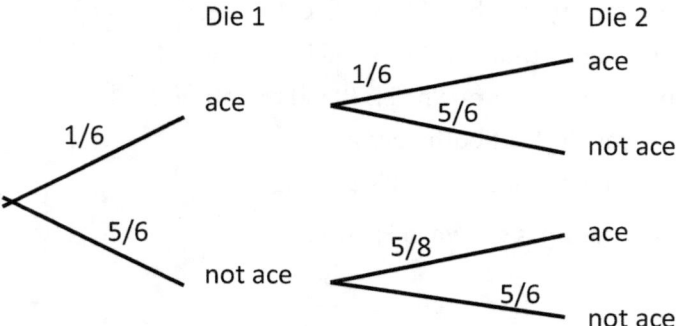

Figure 15. Tree diagram with branch probabilities.

The required probability is

    P(ace then not ace) + P(not ace then ace)

    $= (1/6)(5/6) + (5/6)(1/6) = 10/36$

We note that the falling of die 1 is independent of the falling of die 2. In other words, the 2 dice have no influence on each other, as seen by the fact that both forks leading out of die 1 in the Figure are the same.

Problem 18. 1 die is tossed twice. That is the probability that only 1 ace shows on the 2 throws?

Solution. This problem has the same solution as the foregoing one by substituting "toss 1, toss 2" for "die 1, die 2".

Problem 19. 2 dice are tossed. What is the probability that at least 1 die shows an ace?

Solution 1. The event that "at least 1 die shows an ace" is made up of the 3 incompatible events:

(a)   die 1 shows ace then die 2 shows ace

(b)   die 1 shows ace then die 2 shows not ace

(c)   die 1 shows not ace then die 2 shows ace

By reference to the foregoing Figure, we see that the probabilities of each of these events are

(a) $(1/6)(1/6) = 1/36$

(b) $(1/6)(5/6) = 5/36$

(c) $(5/6)(2/6) = 5/36$

The required probability is.

$1/36 + 5/36 + 5/36 = 11/36.$

Solution 2. The event that "no die shows an ace" is the contrary event of the event that "at least one die shows an ace". The event that "no die shows an ace" is the path

"die 1 shows not ace then die 2 shows not ace"

in the foregoing Figure. This event has probability $(5/6)(5/6)$. Hence the required probability is

$1 - (5/6)(5/6) = 11/36$

which is the same answer as before.

Problem 20. 1 die is tossed twice. What is the probability that at least 1 ace shows on the 2 tosses?

Solution. This problem has the same solution as the foregoing one by substituting "toss 1, toss 2" for "die 1, die 2".

Problem 21. 2 dice are tossed. If they do not show the same face, what is the probability that one die shows an ace?

Solution. The condition that they do not both show the same face limits ourselves to the sub-universe made up of the ordered pairs:

	(1, 2)	(1, 3)	(1, 4)	(1, 5)	(1, 6)
(2, 1)		2, 3	2, 4	2, 5	2, 6
(3, 1)	3, 2		3, 4	3, 5	3, 6
(4, 1)	4, 2	4, 3		4, 5	4, 6
(5, 1)	5, 2	5, 3	5, 4		5, 6
(6, 1)	6, 2	6, 3	6, 4	6, 5	

For example) the ordered pair 6, 3 indicates that "die 1 shows 6 and die 2 shows 3". In this sub-universe there are 30 chances, of which the 10 pairs in parentheses are favorable to the event. Hence the required probability is 10/30 = 1/3.

Problem 22. The year was about 1654. The **Chevalier de Mere** played two different dice games. In the first game, he would bet that he could roll at least one 6 on four rolls of one die. In other words, if a six appeared within the first four rolls of one die, he won the bet. If no six appeared within the first four rolls, he lost the bet. On the average, The Chevalier de Mere made money on the first game.

In the second game, the Chevalier de Mere would bet that he could roll at least one double-6 on 24 rolls of two dice. In other words, if a double-6 appeared within the first four rolls of two dice, he won the bet. If no double-6 appeared within the first four rolls, he lost the bet. On the average, the Chevalier de Mere lost money on the second game.

The Chevalier de Mere noticed that the ratio of the number of tosses to the number of possible outcomes is 4/6, or 2/3, on the first game. He also noticed that the ratio of the number of tosses to the number of possible outcomes is 24/36, or 2/3, on the second game. Because the ratio was the same for both games, he could not understand why his experiences with the two games were different. The Chevalier de Mere appealed to Blaise Pascal, who with Fermat, found the solution.

Solution. Let us discus the first game. A single roll of a die has six possible outcomes, corresponding to the faces that land up. If the die is fair, then the probability of any particular face landing up is 1/6. The four tosses are independent. The probability of not getting a 6 in four trials is $(5/6)^4$, or 0.4823. Thus the probability of getting at least one 6 in four trials is equal to $1 - 0.4823$, or 0.5177. Because the probability of his winning the bet was 51.8%, the Chevalier de Mere had a slight advantage in attempting to roll a 6 in 4 tosses of a single die.

Let us discus the second dice game. In games that involve rolling two dice, the dice are distinguishable. The outcomes (3, 4) and (4, 3) are counted separately. As a result, a roll of two dice has 36 possible outcomes, and the probability of any one of them is 1/36. The four tosses are independent. Hence, the probability of not getting a double-6 in twenty-four trials is $(35/36)^{24}$, or 0.5086. It follows that the probability of getting a double-6 in twenty-four trials is equal to 1 − 0.5086, or 0.4914. Because the probability of his winning the bet was 49.1%, the Chevalier de Mere had a slight disadvantage in attempting to roll a double-6 in 24 tosses of a single die.

The Chevalier de Mere asked, "When one plays with two dice, what is the minimum number of throws on which one can advantageously bet that a double six will turn up?" Pascal gave the solution: 24 throws would be a bad bet; 25, a good one.

Problem 23. The numbers of balls in 2 urns are

	urn 1	urn 2
R (red) balls	a	c
B (black) balls	b	d

One ball is transferred from urn 1 to urn 2, and then a ball is drawn from urn 2. What's the probability that the ball drawn is R (red)?

Solution. The tree diagram is shown in Figure 16.

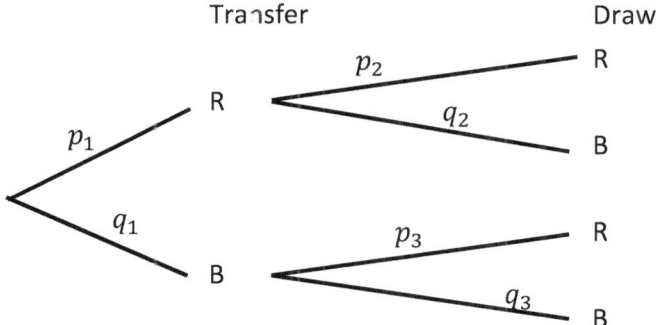

Figure 16. Tree diagram with as yet unknown branch probabilities

The probability that the transferred ball is R is

$$p_1 = \frac{a}{a+b}$$

The transfer of a ball R to urn 2 means that urn 2 now contains (c + 1) balls R and d balls B. Thus the branch probability that the ball drawn is R, given that the ball transferred is R, is

$$p_2 = \frac{c+1}{c+d+1}$$

The probability that the transferred ball is B is

$$q_1 = \frac{b}{a+b}$$

The transfer of a ball B to urn 2 means that urn 2 now contains c balls R and d+1 balls B. Thus the branch probability that the ball drawn is R, given that the ball transferred is R, is

$$p_3 = \frac{c}{c+d+1}$$

The event made up of the path R, then path R, has probability

$$p_1 p_2 = \frac{a}{a+b} \frac{c+1}{c+d+1}$$

and the event made up of the path B, and then path R, has probability

$$q_1 p_3 = \frac{b}{a+b} \frac{c}{c+d+1}$$

These 2 events are incompatible, and their union is the event that the ball drawn is R. Hence the required probability is

$$\frac{a}{a+b} \frac{c+1}{c+d+1} + \frac{b}{a+b} \frac{c}{c+d+1}$$

Problem 24. There are 5 coins in purse 1. Of these, 4 are S (silver) and 1 G (gold). There are 5 coins in purse 2, all of which are S (silver). Suppose that 4 coins are drawn haphazardly from purse 1 and put into purse 2. Then 4 coins are drawn haphazardly from purse 2 and put into purse 1. A person may now pick whichever purse he pleases. Which purse should he pick?

Solution. At the end, each purse contains the same number of coins, so he ought to pick the purse that has the greater probability to contain G (the gold coin). Now G can only be in purse 2 provided the following path is taken.

p_1	G among the 4 coins drawn from purse 1	q_2	G not among the 4 coins drawn from purse 2
$\longrightarrow$		$\longrightarrow$	

Instead of directly computing the probability of the event that G was among the 4 coins drawn from purse 12 let us first compute the probability of the contrary event. The contrary event is made up of one path; namely

4/5	not G	3/4	not G	2/3	not G	1/2	not G
$\longrightarrow$		$\longrightarrow$		$\longrightarrow$		$\longrightarrow$	

which has probability

$$q_1 = \frac{4}{5}\frac{3}{4}\frac{2}{3}\frac{1}{2} = \frac{1}{5}$$

Therefore the probability of the event that G was among the 4 coins drawn from purse 1 is

$$p_1 = 1 - q_1 = 1 - \frac{1}{5} = \frac{4}{5}$$

Given that G was among the 4 coins put into purse (so purse 2 now contains 8 S and 1 G, the event that G was not among the 4 coins drawn from purse 2 is made up of one path) namely

8/9	not G	7/8	not G	6/7	not G	5/6	not G
$\longrightarrow$		$\longrightarrow$		$\longrightarrow$		$\longrightarrow$	

which has probability

$$q_2 = \frac{8}{9}\frac{7}{8}\frac{6}{7}\frac{5}{6} = \frac{5}{9}$$

Hence the probability that G is in purse 2 is

$$p_1 q_1 = \frac{4}{5}\frac{5}{9} = \frac{4}{9}$$

so the person should pick purse 1.

CHAPTER 15. GEOMETRIC PROBABILITY

Blaise Pascal wrote: Of these two Infinites of science, that of greatness is the most palpable, and hence few persons have pretended to know all things. But the infinitely little is the least obvious. Philosophers have much oftener claimed to have reached it, and it is here they have all stumbled.

15.1 Geometric probability

Problems of geometric probability have been influential in the modern development of probability theory. They will provide us with some instructive and interesting examples. Moreover such problems find many important applications in science and technology. In the preceding chapters, we have considered in detail stochastic phenomena where there are a finite number of possible results, or cases. This situation is different for geometric stochastic phenomena where the possible results, or cases, from an infinite and continuous set.

Stochastic phenomena of a geometrical nature are ones for which the possible results, or cases, are possible geometrical positions. For example, the possible geometrical positions may be the possible positions of a point on a line. In another example, the possible results may be the possible positions of a point in a plane area. In still another example, the cases may be the possible positions of a straight line in space.

15.2 A definition of geometric probability

Let us first look at a simple example of geometric probability, namely an example of points on a line. Let AB be the straight line in question. We now suppose that the point M has been placed haphazardly on the line AB. See Figure.

A   M   B

Figure 1: Haphazard point M on line AB.

 We now ask: What is the probability that point M occupies a determined position on segment AB?

This question should be distinguished from questions about probability that we have considered up to now. Now the number of cases is infinite. The point M can vary in a continuous manner between A and B. As a result, our definition of probability as the ratio of the number of favorable cases to the number of all cases is no longer applicable. We must therefore seek a new definition of probability that will be applicable to geometrical phenomena. Let us mark an arbitrary segment PQ on the line AB. See Figure.

| A | P | Q | B |

Figure 2. Arbitrary segment PQ on the line AB.

The totality of admissible points are on the line AB. The favorable points are on the segment PQ. The segment PQ is perfectly arbitrary. It can be any segment that we choose. In other words, the points P and Q may be any arbitrary points on the line AB.

Moreover we allow the 2 extremes; namely, where

(1) the segment PQ coincides with the entire line AB, and

(2) the segment PQ is only a point (that is, the case where the points P and Q coincide).

Let us now adopt the following definition of probability for this example.

DEFINITION. The probability that the haphazard point M falls on the segment PQ of the line AB is equal to that ratio of the length of PQ to the length of AB.

The segment PQ is an event. According to our definition, the probability of this event is the ratio

$$\text{probability} = \frac{\text{length of PQ}}{\text{length of AB}}$$

15.3 Objections to the definition

Let us now illustrate this definition. A haphazard number x falls on the straight line between 2 and 5. According to our definition, that probability that x falls on the segment between 2 and 3 is

$$\text{probability} = \frac{\text{length between 2 and 3}}{\text{length between 2 and 5}} = \frac{3-2}{5-2} = \frac{1}{3} = 0.3333\ldots$$

As a second illustration, let us consider the square of x; that is, x^2. When x is between 2 and 5, x^2 is between 4 and 25. When x is between 2 and 3, x^2 is between 4 and 9. Thus the probability that x^2 falls between 4 and 9, given that it must fall somewhere between 4 and 25, is

$$\text{probability} = \frac{9-4}{25-4} = \frac{5}{21} = 0.2380\ldots$$

As a third illustration, let us consider 1/x. When x is between 2 and 5, then 1/x is between 1/5 and 1/2. When x is between 2 and 3, then 1/x is between 1/3 and 1/2. The probability that 1/x falls between 1/3 and 1/2, given that it must fall somewhere between 1/5 and 1/2, is

$$\text{probability} = \frac{\frac{1}{2}-\frac{1}{3}}{\frac{1}{2}-\frac{1}{5}} = \frac{\frac{1}{6}}{\frac{3}{10}} = \frac{5}{9} = 0.5555\ldots$$

We might conclude that the definition is unacceptable, because it gives different answers when x is replaced by a simple function of such as x^2 or 1/x. By the proper choice of a function of x, it is possible to make the probability have any arbitrary value.

This objection to the definition is valid, but its practical importance is not very great. From a practical point of view, this objection serves to warn us of the mistakes that can result from a poor choice of the variable used in this definition of probability.

This definition of probability represents a probability model. To apply this model, we must first choose the proper variable to use in this definition. Whenever we deal with concrete problems and not abstract ones, this choice of variable is almost always imposed in some way by the conditions of the problem. Thus this choice should be made by utilizing our knowledge of the connection of the problem with the real world. Then some arbitrary different choice, obtained by means of a mathematical change of variables, would destroy this connection with reality, and so would be only a mathematical exercise. We must always establish that the assumption underlying the choice of variable is in accord with empirically observed facts. Only then can we make use of practical applications of this probability model.

In the above illustrations, we would study the connection of the problem with the empirical world. For one particular application, we might conclude that the proper choice of variable to use in the definition is x, not x^2, nor 1/x, and so the desired probability would be 1/3, and neither 5/21 nor 1/20. For some other application, we might conclude that the proper choice of variable to use in the probability definition is x^2, not x, nor 1/x. Then the desired probability would be 5/21, and neither 1/3 nor 1/20. Always our choice must depend upon our first establishing which choice is in accord with the empirical facts.

15.4 Bertrand's paradox

> Blaise Pascal wrote: The heart has its reasons, which reason does not know. We feel it in a thousand things. It is the heart which experiences God, and not reason. This, then, is faith: God felt by the heart, not by reason.

As an example of the proper choice of variable, let us consider a famous problem introduced by Joseph Bertrand (Calcul des Probabilitea, Paris, 1889, p.4) and now called Bertrand's paradox.

Let us now explain **Bertrand's paradox**.

A chord is drawn haphazardly in a circle. Consider the event that the length of the chord exceeds the length of the side of the inscribed equilateral triangle. What is the probability of this event?

This problem is cited by Bertrand as a problem whose conditions are incomplete. That is, the conditions do not specify which variable should be chosen for the probability definition. More specifically, the concept of drawing a chord haphazardly is left undefined in the conditions of the problem. We can take advantage of this ambiguity by choosing different variables to use in our definition of probability. Each different choice will result in a different probability. As an illustration, let us specify 3 different choices.

Choice 1. The circle with center A and the haphazardly drawn chord CD are shown in Figure 3.

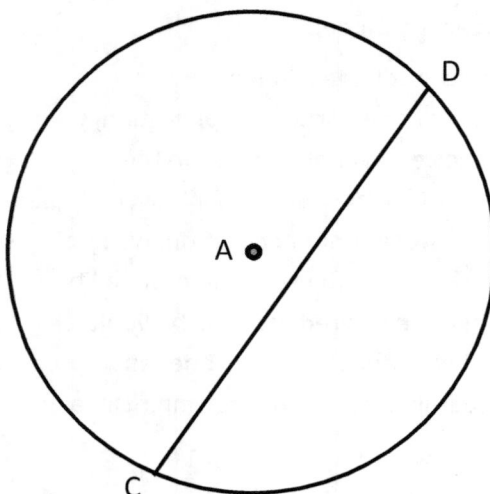

Figure 3. Circle A and chord CD.

Let us now bisect the chord CD by the radius AB and let M denote the point of intersection, as shown in Figure 4.

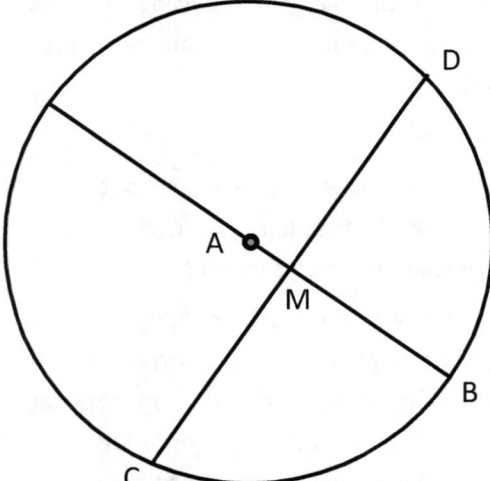

Figure 4. Radius AB and point of intersection M.

Now we want to compare the length of the chord CD with the length of the side of the inscribed equilateral triangle. Locate the equilateral triangle EFG so that one side EF is parallel to the chord CD. From plane geometry, t can be shown that the point of intersection Q of the side EF with the radius AB s at the midpoint of radius AB. See Figure 5.

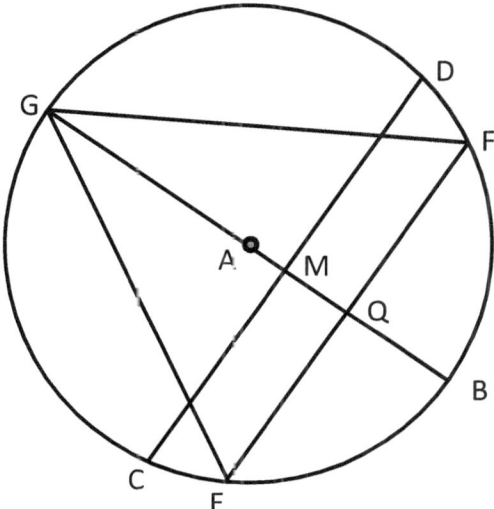

Figure 5. Equilateral triangle EFG and point of intersection Q (at midpoint of radius AB).

From the Figure we see that the length of chord CD exceeds the length of side EF when M falls on the line segment AQ. In other words, points on the line segment AB between A and Q are favorable points for the event. Thus if we choose the point M to have an equal chance anywhere on the radius AB, it follows that the probability of the event is

$$\text{probability} = \frac{\text{length of AQ}}{\text{length of AB}} = \frac{1}{2}$$

Choice 2.

Again let the haphazardly drawn chord be CD. Draw the tangent AB to the circle at point C. See Figure 6.

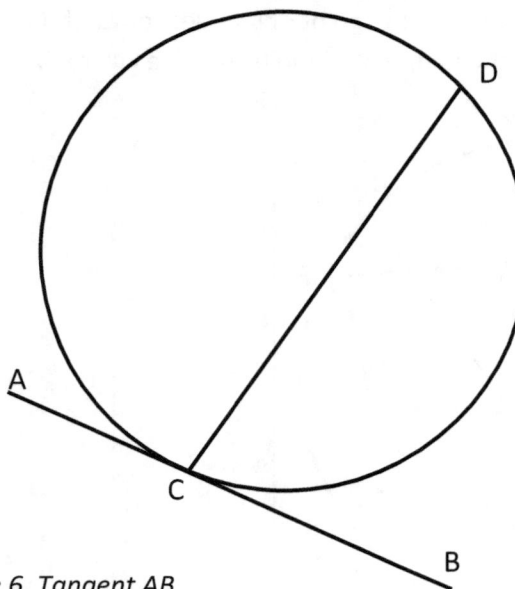

Figure 6. Tangent AB

Now we want to compare the length of the chord CD with the length of the side of the inscribed equilateral triangle. Locate the equilateral triangle CEF so that C is one of its vertices. See Figure 7.

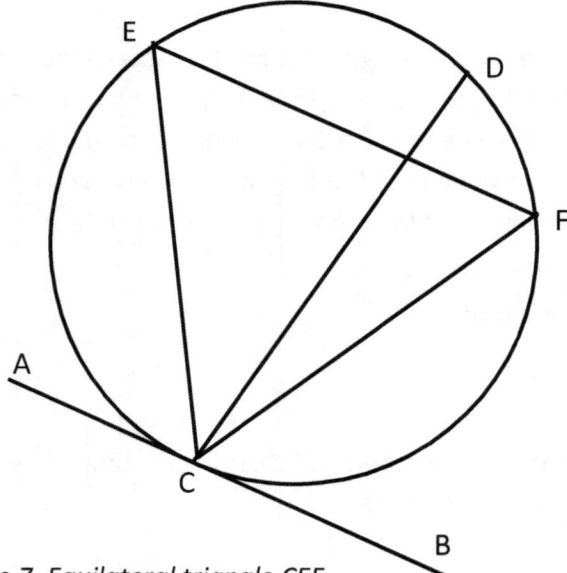

Figure 7. Equilateral triangle CEF.

From plane geometry we know that angle ACE is 60° and that angle ACF is 120°. Thus the length of chord CD exceeds the length of the side of the equilateral triangle, provided angle ACD is between 60° and 180°. Let us choose the angle ACD to be the variable x of the problem, so that x has equal chances anywhere from 0 to 180°. The points favorable to the event lie on the sub-segment from 60° to 120°. For this choice, the probability is

$$\text{probability} = \frac{120 - 60}{180 - 0} = \frac{60}{180} = \frac{1}{3}$$

Pascal wrote: We naturally believe ourselves far more capable of reaching the center of things than of embracing their circumference. The visible extent of the world visibly exceeds us; but as we exceed little things, we think ourselves more capable of knowing them. And yet we need no less capacity for attaining the Nothing than the All. Infinite capacity is required for both, and it seems to me that whoever shall have understood the ultimate principles of being might also attain to the knowledge of the Infinite. The one depends on the other, and one leads to the other. These extremes meet and reunite by force of distance, and find each other in God, and in God alone.

Let us then take our compass; we are something, and we are not everything. The nature of our existence hides from us the knowledge of first beginnings which are born of the Nothing; and the littleness of our being conceals from us the sight of the Infinite. Our intellect holds the same position in the world of thought as our body occupies in the expanse of nature.

Let us therefore not look for certainty and stability. Our reason is always deceived by fickle shadows; nothing can fix the finite between the two Infinites, which both enclose and fly from it. If man made himself the first object of study, he would see how incapable he is of going further. How can a part know the whole? But he may perhaps aspire to know at least the parts to which he bears some proportion. But the parts of the world are all so related and linked to one another, that I believe it impossible to know one without the other and without the whole.

CHAPTER 16. PROBLEMS

Blaise Pascal wrote: The struggle alone pleases us, not the victory. ...It is the same in play, and the same in the search for truth. In disputes we like to see the clash of opinions, but not at all to contemplate truth when found. ...So in the passions, there is pleasure in seeing the collision of two contraries; but when one acquires the mastery, it becomes only brutality. We never seek things for themselves, but for the search.

16.1 Permutations

We are often faced with the problem of combining things. If the order of the things does not matter, then the word "combination" is used. If order does matter, then the word "permutation" is used.

Let us first talk about permutations. The two types are "permutation with repetition allowed" and "permutation with repetition not allowed."

Let us discuss permutations with repetition. We want to choose a 3-digit number. Each digit can be anyone of the 10 digits 0, 1, 2, 3, 4, 5, 6, 7, 8, 9. For example, the chosen number might be 339. The number of permutations is 10 × 10 × 10 = 1000. In other words, there are 1000 3-digit numbers from 000 to 999. In case we have n things and want to choose r things, there are n possibilities for the first choice, and then there are n possibilities for the second choice, and so on, up to n possibilities for the choice r. Thus the number of permutations is

$$n \times n \times \; ... \; (r \; times) = n^r$$

Let us discuss permutations without repetition. Again we have n things from which to choose, and we choose r of them. However, once a thing is chosen, it cannot be chosen again. For example, if we are given the three letters A, B, and C, then there are 6 permutations of these letters; namely, ABC, ACB, BAC, BCA, CAB, and CBA.

A teacher has 20 students in her class, and she wants to choose 3 students to work in the library. There are 20 ways to choose the first student. Having selected the first student, there are 19 ways left to choose the second student. We see that there are 20 × 19 ways of choosing the first 2 students. There are 8 ways left to choose the third student. We see that there are 20 × 19 × 18 = 6840 ways in which to choose the 3 students.

We want to choose a 3-digit number, but with no digit repeated. After choosing the digit 4, we cannot choose it again. In the first choice, we have 10 possibilities. In the next choice we have 9 possibilities. For the third choice we have 8 possibilities. The total number of permutations is 10 × 9 × 8 = 420.

In other words, there are 720 different 3-digit numbers with no digit used more than once. At his point, we want to introduce the "factorial function." The exclamation point "!" is the symbol for factorial. It means that we multiply a series of descending natural numbers. For example,

$n! = n \times (n-1) \times (n-2) \times ... \times 3 \times 2 \times 1$

$10! = 10 \times 9 \times 8 \times 7 \times 6 \times 5 \times 4 \times 3 \times 2 \times 1$

It follows that 1! = 1. Mathematical convention is that 0! = 1. The equation for the number of permutations of n objects taken n at a time is

$$P(n, n) = n!$$

(Be careful to note that the same symbol P is used both for probability and for permutation.) In selecting a 3-digit number, we have 10 × 9 × 8 permutations. In other words, we stop multiplying after 8. In other words, we compute

$$\frac{10!}{7!} = \frac{10 \times 9 \times 8 \times 7 \times 6 \times 5 \times 4 \times 3 \times 2 \times 1}{7 \times 6 \times 5 \times 4 \times 3 \times 2 \times 1} = 10 \times 8 \times 9$$

The formula for the number of permutations of n objects taken r at a time is

$$P(n, r) = \frac{n!}{(n-r)!}$$

16.2 Combinations

A player is dealt a poker hand consisting of five spades in the order 5, K, 7, 8, 4. The player would be equally happy to receive the same cards in any order. In other words, the player is interested only in the membership of the set and not in any particular sequence of the membership. The word combination is used to cover such sets. The number of combinations of n things taken r at a time is designated by

$$\binom{n}{r} = \frac{n!}{(n-r)!\,(r)!}$$

Suppose 7 students want to go on a field trip. However there is room only for 3 of them. The problem is to find in how many ways can 3 students be chosen from the group of 7. In other words, we want the number of combinations of 7 objects taken 3 at a time. It is

$$\binom{7}{3} = \frac{7!}{(7-3)!\,(3)!} = \frac{7!}{4! \times 3!} = \frac{7 \times 6 \times 5}{3 \times 2 \times 1} = 7 \times 5 = 35$$

Let us find the probability of getting any specified hand at a game of cards. Suppose it's desired to determine the chances of getting a full house at poker; that is, of a pair with 3 of a kind. Let us first find the number of combinations in which a pair can occur. Suppose it is a pair of 5's. The number of different pairs that can be made out of four 5's is equal to the number of combinations of 4 objects taken 2 at a time. It is

$$\binom{4}{2} = \frac{4!}{(4-2)! \times (2!)} = \frac{4 \times 3 \times 2 \times 1}{2 \times 1 \times 2 \times 1} = 6$$

Hence there are 6 ways in which a pair of 5's can be obtained. Since the same reasoning holds true for all 13 numbers, there are 6 × 13 = 78 combinations of a pair. Using the same principles, there are

$$\binom{4}{3} = \frac{4!}{(4-3)! \times (3!)} = \frac{4 \times 3 \times 2 \times 1}{1 \times 3 \times 2 \times 1} = 4$$

ways of getting 3 kings, or 3 of any other rank. The pair was one of the 13 ranks, so there are only 12 left Hence the number of ways of getting 3 of a kind is equal to 4 × 12= 48. Each 3 of a kind can be used with each pair. Thus there are 6 × 13 × 4 × 12 = 78 × 48 = 3,744 ways of the occurrence of a full house.

The number possible outcomes is the total number of different poker hands, which is the number of combinations of 52 cards taken 5 at a time, given by

$$\binom{52}{5} = \frac{52!}{(52-5)!\,(5)!} = \frac{52!}{47! \times 5!} = \frac{52 \times 51 \times 50 \times 49 \times 48}{5 \times 4 \times 3 \times 2 \times 1}$$

The probability of a full house is equal to the 78 × 48 favorable outcomes divided by the total number of possible outcomes. It is

$$p = \frac{6 \times 13 \times 4 \times 12}{\left[\dfrac{52 \times 51 \times 50 \times 49 \times 48}{5 \times 4 \times 3 \times 2 \times 1}\right]} = \frac{6}{4165} = 0.001440576$$

16.3 Pascal's Triangle

Pascal writes: The word Combination has been taken in many different senses, so that, in order to remove the ambiguity, I am obliged to speak as I intend it. When among many things we give the choice of a certain number, all the ways of taking from them as many as is permitted among all those which are presented, are called here the different combinations.

For example, if from four things expressed by these four letters, A, B, C, D, we permit to take from them, for example, any two, all the ways of taking from them two different in the four which are proposed, are called Combinations. Thus we will find, by experience, that there are six different ways of choosing two in four; we are able to take A and B, A and C or A and D, or B and C, or B and D, or C and D. Or thus:

The number of combinations of 0 in 4 is 1.
The number of combinations of 1 in 4 is 4.
The number of combinations of 2 in 4 is 6.
The number of combinations of 3 in 4 is 4.
The number of combinations of 4 in 4 is 1.

Another name of the arithmetic triangle is Pascal's Triangle. The top part of the infinite Pascal's triangle is

Row									
0					1				
1				1		1			
2			1		2		1		
3		1		3		3		1	
4	1		4		6		4		1
5	1	5		10		10		5	1
6	1	6	15		20		15	6	1
7	1	7	21	35		35	21	7	1

Its rows are numbered starting at 0. The left column identifies row 0, row 1, row 2, etc.

The top entry on Pascal's Triangle is the number 1. The edges on both sides are also made up of the number 1. Any other number is the sum of the two adjacent numbers above it. To construct the triangle, start with 1 at the top and on both edges. Then go down the rows by adding adjacent entries on the row above. For example, add the 1 and 1 on row 1 to get the 2 on row 2. Add the 1 and 2 on row 2 to get the first 3 on row 3. Add the 2 and 1 on row 2 to get the second 3 on row 3. Pascal's triangle is symmetrical, in the sense that the numbers on the left side match the numbers on the right side.

The entries on row n are identified by the index numbers

0, 1, 2, ... , n–1, n.

For example, entries 0, 1, 2, 3 on row 3 are respectively 1, 3, 3, 1. The entry 0 on row 0 is 1.

The sum on the entries on row 1 is 1+1=2. The sum on the entries on row 2 is 1+2+1=4. The sum on the entries on row 3 is 1+3+3+1=8. The sum on the entries on row 4 is 1+4+6+4+1=16. The sum doubles each time; it is a power of 2.

Pascal states,

> Whence it is seen that the combinations of 2 in 4 are formed by the combinations of 1 in 3, and of 2 in 3; and hence that the number of combinations of 2 in 4 equals that of 1 in 3, and of 2 in 3.

In mathematical notation, Pascal's statement is

$$\binom{4}{2} = \binom{3}{1} + \binom{3}{2} \text{ or } 6=3+3$$

Pascal states,

> We will show the same thing in all the other examples, as:
>
> The number of combinations of 29 in 40;
>
> and the number of combinations of 30 in 40:
>
> equals the number of combinations of 30 in 41.

In mathematical notation, Pascal's statement is

$$\binom{40}{29} + \binom{40}{30} = \binom{41}{30}$$

or 2311801440+847560528=3159461968

Pascal's triangle in terms of the symbols for combinations is

$$\binom{0}{0}$$

$$\binom{1}{0} \qquad \binom{1}{1}$$

$$\binom{2}{0} \qquad \binom{2}{1} \qquad \binom{2}{2}$$

$$\binom{3}{0} \qquad \binom{3}{0} \qquad \binom{3}{0} \qquad \binom{3}{0}$$

$$\binom{4}{0} \qquad \binom{4}{1} \qquad \binom{4}{2} \qquad \binom{4}{3} \qquad \binom{4}{4}$$

For example, row 4, entry 2 in Pascal's Triangle is

$$\binom{4}{2} = \frac{4!}{2\,(4-2)!} = \frac{4!}{2!\,(2)!} = \frac{4 \times 3 \times 2 \times 1}{2 \times 1 \times 2 \times 1} = 6$$

Pascal's triangle gives the coefficients in the binomial expansion:

Binomial Expansion	Pascal's triangle

$(x + 1)^0 = \mathbf{1}$ 1

$(x + 1)^1 = \mathbf{1}x + \mathbf{1}$ 1   1

$(x + 1)^2 = \mathbf{1}x^2 + \mathbf{2}x + \mathbf{1}$ 1   2   1

$(x + 1)^3 = \mathbf{1}x^3 + \mathbf{3}x^2 + \mathbf{3}x + \mathbf{1}$ 1   3   3   1

$(x + 1)^4 = \mathbf{1}x^4 + \mathbf{4}x^3 + \mathbf{6}x^2 + \mathbf{4}x + \mathbf{1}$ 1   4   6   4   1

For this reason, the entries of Pascal's triangle are called binomial coefficients.

Let us look at Pascal's triangle. It shows you how many combinations of objects are possible. You have 4 objects: an apple, an orange, a peach, and a fig.

On Pascal's triangle, the top row is row 0. Row 4 pertains to the 4 objects in question. It contains the entries 1, 4, 6, 4, 1.

The first entry 1 gives the number of ways of choosing 0 of the four objects, ignoring the order of objects in the in the selection.

The second entry 4 gives the number of ways of choosing 1 of the four objects, ignoring the order of objects in the selection.

The third entry 6 gives the number of ways of choosing 2 of the four objects, ignoring the order of objects in the selection.

The fourth entry 4 gives the number of ways of choosing 3 of the four objects, ignoring the order of objects in the selection.

The fifth entry 1 gives the number of ways of choosing 4 of the four objects, ignoring the order of objects in the selection.

The combinations of heads and tails are shown by Pascal's Triangle. In 3 tosses of a coin, there is 1 instance of three heads (HHH), 3 instances of two heads and one tail (HHT, HTH, THH), 3 instances of one head and two tails (HTT, THT, TTH), and 1 instance of three tails (TTT).

This pattern "1, 3, 3, 1" gives the entries in the row 3 of Pascal's triangle. In 3 tosses of coin, what is the probability of getting just one head? The number of possible results is 1+3+3+1 = 8. Of these, 3 give exactly just one head. The probability is 3/8.

16.4 Two birthdays on the same day

Among a group of 24 people, what are the chances of 2 birthdays on the same day? There are 365 days in the year on which their birthdays may fall. It would seem the chances would be small. However, the chances are slightly better than even. Let us find the probability that each of the 24 persons has a different birthday. The first person in the group can have any birthday. Because one of the 365 days has been used by the first person, the chances are 364 out of 365 that the birthday of the second person is different. The probability of this event is 364/365. Now, what are the chances that the third person has a different birthday? The probability is 363/365, because 2 days of the year have been excluded. It follow that the probabilities that each of the remaining persons will have a different birthday are 361/365, 360/365, and so on to the twenty-fourth person, for whom the probability is 342/365. The probability that any two persons do not have the same birthdays is

$$\left(\frac{364}{365}\right) \times \left(\frac{363}{365}\right) \times \cdots \times \left(\frac{342}{365}\right) = 0.46$$

In other words, there are 46 chances in 100 that no two of the group of 24 people have birthdays on the same date. Thus the probability that any two persons do have the same birthdays is 1-0.46 = 0.54.

16.5 General problems

Problem 1. A letter is chosen at random out of each of the words "tinsel" and "silent." What is the probability that they are the same letter? What is the probability, if it is known that they are either both consonants or both vowels?

Solution. Whatever letter is chosen out of "tinsel," there is a probability 1/6 of getting the same letter out of" silent," so 1/6 is the required probability. If it is known that both are consonants or both are vowels, the probability of getting a vowel from "tinsel" is 1/3, and then the probability of getting the same vowel from "silent" is 1/2. Also the probability of getting a consonant from "tinsel" is 2/3, and the probability of getting the same consonant from "silent" is 1/4. Hence the required probability is

(1/3)(1/2) + (2/3)(1/4) = 1/3 .

Problem 2. 120 men are formed at random into a solid rectangle of 10 by 12 men; each side of the rectangle is equally probable to be in front. What is the probability that an assigned man is in front?

Solution.

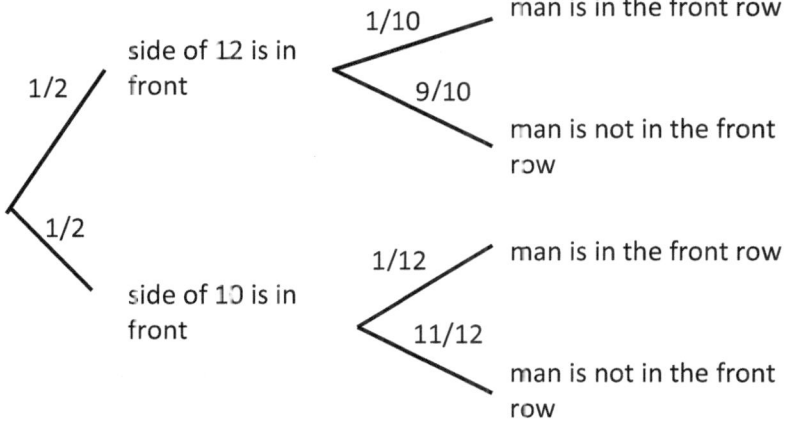

The required probability is

(1/2)(1/10) + (1/2)(1/12) = 11/120 .

Problem 3. 2 numbers are chosen at random. Find the probability that their sum is even.

Solution. The probability that both are even is (1/2)(1/2) = 1/4. The probability that both are odd is (1/2)(1/2) = 1/4 The required probability is thus 1/4 + 1/4 = 1/2.

Problem 4. 2 letters are taken at random out of "cocoa" and 2 out of "cocoon." Find the probability that the 4 letters should be all different.

Solution.

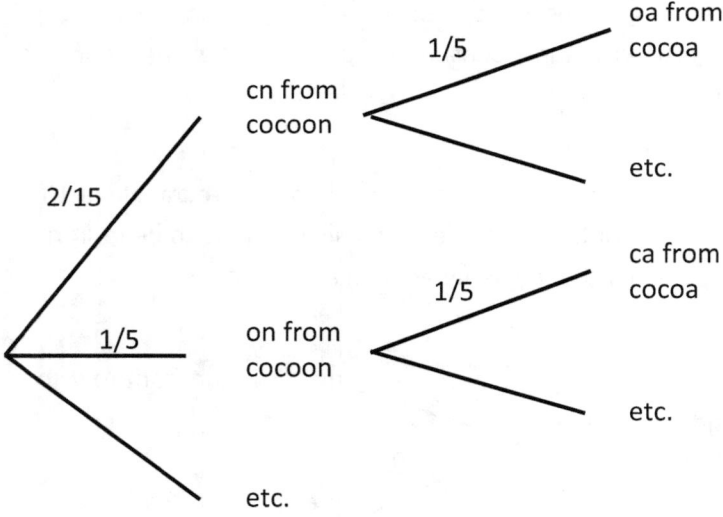

The required probability is

(2/15)(1/5) + (1/5)(1/5) = 1/15 .

Problem 5. At random, a letter is taken out of "assistant" and a letter out of "statistics." What is the probability that they are the same letter?

Solution.

Letter combination	Number of ways
s s	3×3 = 9
t t	2×3 = 6
a a	2×1 = 2
i i	1×2 = 2

Sum = 19 = number of favorable ways

The number of all ways is 9×10 = 90. The required probability is thus 19/90.

Problem 6. An urn contains 4 black balls and 1 red ball. People draw the balls out in success on. They receive a dollar for every ball they draw, until they draw the red one. What is their expectation?

Solution. The red ball is equally probable to be drawn 1st, 2nd, 3rd, 4th, or 5th, which means that it is equally probable that the people receive 0, 1, 2, 3, or 4 dollars. Their expectation is thus (1/5)(0+1+2+3+4)=2 dollars.

Problem 7. 1 domino is drawn at random out of a set of dominoes (numbered from 1, 1 to 6, 6). At the same time a pair of dice is tossed.

(1) What is the probability that the numbers on the dice will be the same as those on the domino?

(2) What is the probability that they will have at least 1 number in common?

Solution (1). There are 21 dominoes in the set. Whatever the dice show, there is one domino that has the same numbers. Thus the required probability is 1/21.

Solution (2).

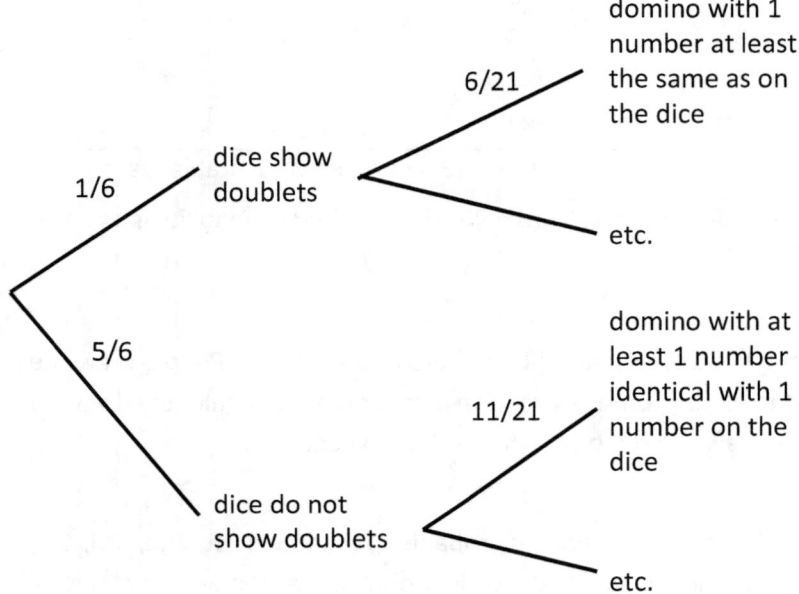

dice show doublets — 1/6

6/21 — domino with 1 number at least the same as on the dice

etc.

5/6

dice do not show doublets

11/21 — domino with at least 1 number identical with 1 number on the dice

etc.

Thus the required probability is (1/6)(6/21)+(5/6)(11/21)=61/126 .

Problem 8. A box contains 10 pairs of gloves. Person A draws out a single glove. Then person B draws one. Then A draws a second; then B draws a second. (1) What is the probability that A draws a pair? (2) What is the probability that B draws a pair? (3) What is the probability that neither draws a pair?

Solution.

A will draw a pair provided

1.   B does not draw the mate to A's first draw.
2.   A draws the mate to his first draw.

The probability that A draws a pair is

(18/19)(1/18) = 1/19

B will draw a pair provided

1.   B does not draw the mate to A's first draw.
2.   A does not draw the mate to B's first draw.
3.   B does draw the mate to his first draw.

The probability that B draws a pair is

$(18/19)(17/18)(1/17) = 1/19$.

(Note: We see that A's and B's probabilities are equal.)

Both will draw a pair provided

1. B does not draw the mate to A's first draw.

2. A draws the mate to his first draw.

3. B draws the mate to his first draw.

The probability that both draw a pair is

$(18/19)(1/18)(1/17) = 1/323$.

The probability that at least 1 draws a pair is

$$\frac{1}{19} + \frac{1}{19} - \frac{1}{323} = \frac{33}{323}$$

The probability that neither draws a pair is thus

$$1 - \frac{33}{323} = \frac{290}{323}$$

Problem 9. In how many ways can a pack of 52 cards be dealt to 13 players, 4 cards to each player, so that every player can have a card of each suit?

Solution. The factorial of a non-negative integer n, denoted by n!, is the product of all positive integers less than or equal to n. For example, $4!=4\times3\times2\times1=24$. The value of 0! is 1. Each suit can be dealt in 13! ways. Thus all the suits can be dealt in $(13!)^4$ ways.

Problem 10. In how many ways can a pack of 52 cards be dealt to 13 players, 4 cards to each player, so that one player can have a card of each suit, and no one else have cards of more than 1 suit.

Solution. The one player can be selected in 13 ways. His cards can be selected in 13^4 ways. Then there will remain 12 cards in each suit. The 12 cards of a given suit can be made into 3 indifferent parcels of 4 cards each in

$$\frac{1}{3!} \frac{12!}{4!4!4!}$$

ways, and all the 48 cards into 12 parcels in

$$\left(\frac{1}{3!}\frac{12!}{4!4!4!}\right)^4$$

ways. Thus we get 12 parcels each containing 4 cards of 1 suit. These can be assigned to the 12 players in 12! ways. Hence the required number of ways is

$$\frac{13^5}{(4!)^{12}}\frac{(12!)^5}{(3!)^4} = \frac{(13!)^5}{(4!)^{12}(3!)^4} = \frac{(13!)^5}{2^{40}\,3^{16}}$$

Problem 11. Suppose that a laboratory stock of 1,000 female rats is made up of:

(1) 950 normal healthy females

(2) 50 healthy females that carry a sex-linked lethal gene.

Accordingly the probabilities that an individual offspring will be female are

$$p_1 = \frac{1}{2} \qquad \text{if the mother is normal}$$

$$p_2 = \frac{2}{3} \qquad \text{if the mother is a carrier}$$

Assume that the elimination of male progeny by the lethal gene does not appreciably affect the total number of live-born rats. A mother rat has eight offspring, all of which are female. What is the probability that the mother rat is a carrier?

Solution.

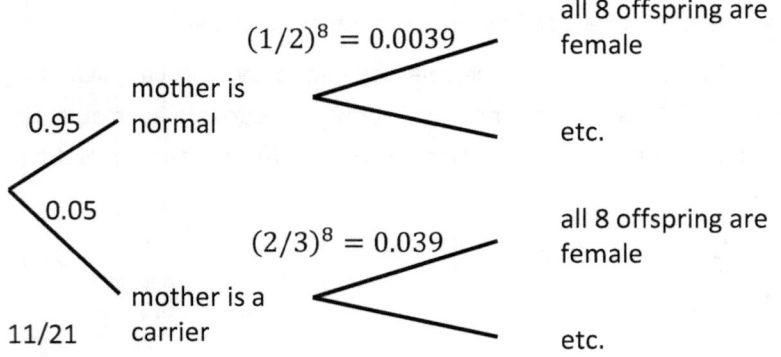

all 8 offspring are female

$(1/2)^8 = 0.0039$

mother is
0.95   normal

etc.

0.05

$(2/3)^8 = 0.039$

all 8 offspring are female

mother is a
11/21   carrier

etc.

The required probability is

$$\frac{(0.05)(0.039)}{(0.95)(0.0039) + (0.05)(0.039)} = 0.345$$

16.6 Probability problems on card hands.

Blaise Pascal wrote: Love knows no limit to its endurance, no end to its trust, no fad ng of its hope; it can outlast any:hing. Love still stands when all else has fa len.

Problem 12. A and B play at cards. A's skill is to B's skill as 3 is to 2. Nothing is known as to whether each has a confederate. If one has and the other has not, the chances of the dishonest player are doubled. If both have confederates, their chances remain the same. Suppose that B wins three successive games. What is the probability that he will win the next game?

Solution. The following 4 hypotheses are equally probable a priori:

1. A only has a confederate
2. Neither has a confederate
3. Both have confederates
4. B only has a confederate

We may treat (2) and (3) together so we have the 3 hypotheses:

1. A has an advantage, with probability 1/4,
2. Neither has an advantage, with probability 1/2
3. B has an advantage, with probability 1/4

Consequently the probabilities of the observed event under these 3 hypotheses are as

$$(1/4)^3 : (2/5)^3 : (4/7)^3$$

A posteriori, the probabilities of the 3 hypotheses are as

$$(1/4)^3 : 2(2/5)^3 : (4/7)^3$$

so the probability that B wins the next game is

$$\frac{(1/4)^4 + 2(2/5)^4 + (4/7)^4}{(1/4)^3 + 2(2/5)^3 + (4/7)^3}$$

Problem 13. What is the probability that a hand of 5 cards has at least 2 aces?

Solution. The probability of no aces is

$$\frac{\binom{4}{0}\binom{48}{5}}{\binom{52}{5}} = \frac{\binom{48}{5}}{\binom{52}{5}}$$

The probability of exactly 1 ace is

$$\frac{\binom{4}{1}\binom{48}{4}}{\binom{52}{5}} = \frac{4\binom{48}{4}}{\binom{52}{5}}$$

Thus the required probability is

$$1 - \frac{\binom{48}{5} + 4\binom{48}{4}}{\binom{52}{5}} = \frac{2257}{54145} = 0.0417$$

Problem 14. What is the probability that a hand of 5 cards has exactly 3 aces?

Solution. The pack of 52 cards is made up of 4 aces and 48 non-aces. The required hand of 5 cards is made up of 3 aces and 2 non-aces. The required probability is therefore

$$\frac{\binom{4}{3}\binom{48}{2}}{\binom{52}{5}} = \frac{94}{54145} = 0.00174$$

Problem 15. What is the probability that a hand of 5 cards has 1 pair?

Solution. The probability of no pair is

$$\frac{\binom{13}{5}\binom{4}{1}^5}{\binom{52}{5}}$$

so the required probability is

$$1 - \frac{\binom{13}{5}\binom{4}{1}^5}{\binom{52}{5}} = \frac{2053}{4165} = 0.493$$

Problem 16. What is the probability that a hand of 5 cards has 2 pairs and no more?

Solution 1.

$$\frac{3\binom{13}{3}\binom{4}{2}\binom{4}{2}\binom{4}{1}}{\binom{52}{5}} = \frac{428\binom{13}{3}}{\binom{52}{5}} = \frac{198}{4165}$$

Solution 2.

$$\frac{\binom{13}{2}\binom{4}{2}\binom{4}{2}\binom{11}{1}\binom{4}{1}}{\binom{52}{5}} = \frac{198}{4165}$$

Problem 17. What is the probability that a hand of 5 cards has 3 of a kind, and no more?

Solution 1.

$$\frac{3\binom{13}{3}\binom{4}{3}\binom{4}{1}\binom{4}{1}}{\binom{52}{5}} = \frac{88}{4165}$$

Solution 2.

$$\frac{\binom{13}{1}\binom{4}{3}\binom{12}{2}\binom{4}{1}\binom{4}{1}}{\binom{52}{5}} = \frac{88}{4165}$$

Problem 18. What is the probability that a hand of 6 cards has 4 of a kind, and no more?

Solution 1.

$$\frac{3\binom{13}{3}\binom{4}{4}\binom{4}{1}\binom{4}{1}}{\binom{52}{6}}$$

Solution 2.

$$\frac{\binom{13}{1}\binom{4}{4}\binom{12}{2}\binom{4}{1}\binom{4}{1}}{\binom{52}{5}}$$

Problem 19. What is the probability that a hand of 6 cards has 2 sets of 3 of a kind?

Solution.

$$\frac{3\binom{13}{3}\binom{4}{3}\binom{4}{3}}{\binom{52}{6}}$$

Problem 20. What is the probability that a hand of 6 cards has 4 of a kind and a pair?

Solution 1.

$$\frac{2\binom{13}{2}\binom{4}{4}\binom{4}{2}}{\binom{52}{6}} = \frac{9}{195755}$$

Solution 2.

$$\frac{\binom{13}{1}\binom{4}{4}\binom{12}{1}\binom{4}{2}}{\binom{52}{6}} = \frac{9}{195755}$$

Problem 21. What is the probability that a hand of 5 cards has a sequence of 3 cards and no more (with aces low)?

Solution. There are 2 so-called extreme sequences; namely, the sequence beginning with 1 (i.e., the ace) and the sequence ending with 13 (i.e., the king). An extreme sequence of 3 cards can be selected in

$$2\binom{4}{1}\binom{4}{1}\binom{4}{1}$$

ways, and then there are 52–4×4=36 cards from which the other 2 can be selected. (For example, if the extreme sequence is 1, 2, 3, then the remaining 2 cards must be selected from the cards with values 5, 6, 7, 8, 9, 10, 11, 12, 13, making a total of 36 cards). Thus the number of ways of selecting an extreme sequence of 3 cards and no more is

$$2\binom{4}{1}\binom{4}{1}\binom{4}{1}\binom{36}{2}$$

A medial sequence is a sequence that is not an extreme sequence. A medial sequence can be selected in

$$9\binom{4}{1}\binom{4}{1}\binom{4}{1}$$

ways, and then there are 32 cards from which the other 2 cards can be selected. Thus the number of ways of selecting a medial sequence of 3 cards and no more is

$$9\binom{4}{1}\binom{4}{1}\binom{4}{1}\binom{32}{2}$$

Hence the required probability is

$$\frac{4^3\left[2\binom{36}{2}+9\binom{32}{2}\right]}{\binom{52}{5}}=\frac{7632}{54145}$$

Problem 22. What is the probability that a hand of 4 cards has a sequence of 4 cards (with aces low)?

Solution.

$$\frac{10\binom{4}{1}\binom{4}{1}\binom{4}{1}\binom{4}{1}}{\binom{52}{4}}=\frac{512}{54145}$$

Problem 23. What is the probability that a hand of 4 cards has a sequence of 4 cards (with aces high or low)?

Solution. Both the hands 1, 2, 3, 4 and 11, 12, 13, 1 are now counted, so the required probability is

$$\frac{11\binom{4}{1}\binom{4}{1}\binom{4}{1}\binom{4}{1}}{\binom{52}{4}}$$

Problem 24. Albert holds a hand of 4 cards with 3 aces. Bob holds a hand of 4 cards out of the same pack. What is the probability that Bob has at least 3 of a kind?

Solution. First suppose that Albert's 4th card has been seen and it is not an ace. Let Albert's cards be aaab. Then Bob can have such a hand as

b b b c	in 45 ways
d d d c	in 4×11×44 = 1936 ways
c c c c	in 11 Ways
	Total = 1992 ways

Thus the required probability is

$$\frac{192}{\binom{48}{4}} = \frac{166}{16215}$$

Now suppose that Albert's 4th card has not been seen, so there is a chance that it might be an ace. The probability that it is an ace is 1/49, and then Bob can have such a hand as

d d d c	in 4 -12 .44 2112 ways
c c c c	in 12 ways
	Total = 2124 ways

Therefore Bob's probability becomes

$$\frac{\left(\frac{1}{49} 2124 + \frac{48}{49} 1992\right)}{\binom{48}{4}} = \frac{543}{52969}$$

Problem 25. Suppose that all the 7's have been removed from a pack. What is the probability that a hand of 6 cards should form a sequence (with aces low)?

Solution. The sequences must be from 1 to 6 or from 8 to king. They can be made in

$$2\binom{4}{1}\binom{4}{1}\binom{4}{1}\binom{4}{1}\binom{4}{1}\binom{4}{1}$$

ways, so the required probability is

$$\frac{2\binom{4}{1}\binom{4}{1}\binom{4}{1}\binom{4}{1}\binom{4}{1}\binom{4}{1}}{\binom{48}{6}} = \frac{1024}{1533939}$$

Problem 26. Suppose that all the 10's have been removed from a pack. What is the probability that a hand of 6 cards should form a sequence (with aces low)?

Solution. The sequences must be from 1 to 6, or 2 to 7, or 3 to 8, or 4 to 9, and thus can be made in

$$4 \binom{4}{1} \binom{4}{1} \binom{4}{1} \binom{4}{1} \binom{4}{1} \binom{4}{1}$$

ways, so the required probability is

$$\frac{4 \binom{4}{1} \binom{4}{1} \binom{4}{1} \binom{4}{1} \binom{4}{1} \binom{4}{1}}{\binom{48}{6}} = \frac{2048}{1533939}$$

Problem 27. We see that 3 cards out of a hand of 6 are the 7, 8, 9. What is the probability that the whole hand forms a sequence (with aces low)?

Solution. Eliminating the 3 cards that have been seen, there remain 49 cards. Out of these 49 cards, the other 3 in the hand can be selected in

$$\binom{49}{3}$$

ways. The favorable ways must complete a sequence either from 4, 5, 6, or 7, so the number of favorable ways is

$$4 \binom{4}{1} \binom{4}{1} \binom{4}{1} = 256$$

Hence the required probability is

$$\frac{4 \binom{4}{1} \binom{4}{1} \binom{4}{1}}{\binom{49}{3}} = \frac{32}{2303}$$

Problem 28. We see that 3 cards out of a hard of 6 cards are the 7, 8, 9. What is the probability that it contains 3 pairs?

Solution. A favorable selection is one made up of another 7, 8, 9 out of the remaining 49 cards. This can be made in

$$\binom{3}{1} \binom{3}{1} \binom{3}{1} = 27$$

ways, so the required probability is

$$\frac{\binom{3}{1}\binom{3}{1}\binom{3}{1}}{\binom{49}{3}} = \frac{27}{18424}$$

Problem 29. We see that 3 cards out of a hand of 6 cards are the 7, 8, 9. What is the probability that it contains no pair?

 Solution. Out of the 10 remaining ranks we must select 3 ranks and take 1 card of each rank. Thus the required probability is

$$\frac{\binom{10}{3}\binom{4}{1}\binom{4}{1}\binom{4}{1}}{\binom{49}{3}} = \frac{960}{2303}$$

Problem 30. We see that 3 cards out of a hand of 6 cards are the 7, 8, 10. What is the probability that the whole hand forms a sequence (with aces low)?

Solution. The sequence must be from either 5 or 6 or 7. Thus 3 there are

$$3\binom{4}{1}\binom{4}{1}\binom{4}{1}$$

favorable selections, so the required probability is

$$\frac{3\binom{4}{1}\binom{4}{1}\binom{4}{1}}{\binom{49}{3}} = \frac{24}{2303}$$

Problem 31. Albert and Bob each have a hand of 5 cards out of the same pack. Suppose that Albert's cards form a sequence of 5 beginning from ace. What is the probability that Bob's cards also form a sequence? (Aces are low.)

Solution. Bob's sequence can begin with

ace	in 3×3×3×3×3 ways
2	in 3×3×3×3×4 ways
3	in 3×3×3×4×4 ways
4	in 3×3×4×4×4 ways
5	in 3×4×4×4×4 ways

6	in 4×4×4×4×4 ways
7	in 4×4×4×4×4 ways
8	in 4×4×4×4×4 ways
9	in 4×4×4×4×4 ways

Total = 6439 ways

Therefore the required probability is

$$\frac{6439}{\binom{47}{5}} = \frac{137}{32637}$$

Problem 32. Albert draws a black and a red card from a pack and Bob draws 2 black. What are their respective probabilities of drawing a pair?

Solution. Whatever card Albert draws first, his probability of a second to pair with it is 2/26 = 1/13. Whatever card Bob draws first, his probability of a second to pair with it is 1/25.

Problem 33. In the foregoing problem, show that Bob's probability is the same whether Albert's cards are replaced or not.

Solution. If Albert's cards are not replaced, then Bob must draw 2 cards from 25 black ones, among which there are 12 pairs. His probability is the same as before; namely,

$$\frac{12}{\binom{25}{2}} = \frac{1}{25}$$

Problem 34 In a deal of bridge, what is the probability that a specified player, say North, has at least 2 aces, irrespective of the other 3 players' hands?

Solution. North may have 0, 1, 2, 3, or 4 aces.

The probability that he has no ace is

$$p_0 = \frac{\binom{4}{0}\binom{48}{13}}{\binom{52}{13}}$$

The probability that he has exactly 1 ace is

$$p_1 = \frac{\binom{4}{1}\binom{48}{12}}{\binom{52}{13}}$$

Thus the required probability is

$$1 - p_0 - p_1 = 1 - \frac{\binom{4}{0}\binom{48}{13}}{\binom{52}{13}} - \frac{\binom{4}{1}\binom{48}{12}}{\binom{52}{13}}$$

which is

$$1 - \frac{\binom{48}{13} + 4\binom{48}{12}}{\binom{52}{13}} = 1 - \frac{38 \times 37 \times 11}{49 \times 25 \times 17} = 0.257$$

Problem 35. In a deal of bridge, what is the probability that someone player has at least 2 aces, irrespective of the other 3 player's hands?

Solution. This required probability is not 4 times the probability of the foregoing problem, because the events of the 4 players are not incompatible. The probability that each player has 1 ace is

$$\frac{\binom{4}{1}\binom{48}{12}}{\binom{52}{13}} \frac{\binom{3}{1}\binom{36}{12}}{\binom{39}{13}} \frac{\binom{2}{1}\binom{24}{12}}{\binom{26}{13}} \frac{\binom{1}{1}\binom{12}{12}}{\binom{13}{13}}$$

and the required probability is the complement of this probability; that is, the required probability is

$$1 - \frac{13 \times 13 \times 13}{49 \times 25 \times 17} = 0.895$$

Problem 36. In a deal of bridge, what is the probability that North and no other player has at least 2 aces?

Solution. The following events are incompatible:

1. North and no other player has at least 2 aces (the required event)

2. North and East each have 2 aces

3. North and South each have 2 aces

4. North and West each have 2 aces.

The union of these 4 events is the event that North has at least 2 aces, irrespective of the other 3 players' hands. The probability of this event (we recall from 2 problems above) is 0.0257. Each of events 2, 3, 4 has probability

$$\frac{\binom{4}{2}\binom{48}{11}}{\binom{52}{13}} \frac{\binom{2}{2}\binom{37}{11}}{\binom{39}{13}} \frac{\binom{0}{0}\binom{26}{13}}{\binom{26}{13}} \frac{\binom{0}{0}\binom{13}{13}}{\binom{13}{13}} = 0.02248$$

The required probability is thus

$$0.0257 - 3(0.02248) = 0.190$$

Problem 37. In a deal of bridge, what is the probability that exactly 1 player of the 4 players has at least 2 aces?

Solution. The required event is the union of 4 incompatible events, each with probability 0.190 (found in the foregoing problem). Hence the required probability is 4(0.190) = 0.76 .

> Blaise Pascal wrote: For in fact what is man in nature? A Nothing in comparison with the Infinite, an All in comparison with the Nothing, a mean between nothing and everything Since he is infinitely removed from comprehending the extremes, the end of things and their beginning are hopelessly hidden from him in an impenetrable secret. He is equally incapable of seeing the Nothing from which he was made, and the Infinite in which he is swallowed up.
>
> What will he do then, but perceive the appearance of the middle of things, in an eternal despair of knowing neither their beginning nor their end. All things proceed from the Nothing, and are borne towards the Infinite. Who will follow these marvelous processes? The Author of these wonders understands them. None other can do so.
>
> Through failure to contemplate these Infinites, men have rashly rushed into the examination of nature, as though they bore some proportion to her. It is strange that they have wished to understand the beginnings of things, and thence to arrive at the knowledge of the whole, with a presumption as infinite as their object. For surely this design cannot be formed without presumption or without a capacity infinite like nature.

16.7 Related books

Available from Amazon.com and other retail outlets.

An introduction to Infinitely Many Variates (Griffin's Statistical Monographs: No. 6) by Enders A. Robinson (1959). [Review by Peter Whittle, professor of mathematics at the University of Cambridge, Cambridge, England, in the *Journal of the Royal Statistical Society*, Series A. Vol. 123, Part 4, 1960: "It is remarkable that so much material can be covered in 100 small format pages, and even more remarkable that the proofs are given in fair completeness, and that, considering its matter, the book is fairly readable. There is no doubt that Dr. Robinson has achieved a tour de force in covering this wide field with an account that is at once brief, lucid, and informative, and that many will find his work useful both as a quick introduction and as a rapid reference."]

Random Wavelets and Cybernetic Systems (Griffin's Statistical Monographs: No. 9) by Enders A. Robinson (1962). [Referred to in the book *Fourier Series and Integrals*, the standard mathematics advanced text by H. Dym and H. P. McKean, 1972: "Minimum-phase filters have the biggest power. This beautiful fact is due to Robinson; it does not seem to have found its way into the purely mathematical literature."]

Statistical Communication and Detection with Special Reference to Digital Data Processing of Radar and Seismic Signals, by Enders A Robinson (1967). [Review by R. Barrett in *Electronic Engineering*, December 1967: "This book provides a most absorbing discussion of the use of a digital computer as a data-processing device for processing signals in noise, in the fields of both radar and seismology. The similarities and points of difference in the two fields are well covered and a real effort is made to develop a relationship between the mathematical model and the physical system which it represents."]

Multichannel Time Series Analysis with Digital Computer Programs by Enders A. Robinson (1967). [Review by Ulf Grenander, Professor of Mathematics at Brown University and also at the Royal Institute of Technology of Stockholm, Sweden, in *The Quarterly of Applied Mathematics*, vol. 26, January 1969: "The present book deals with the computational aspects of time series analysis. The style of the book is clear, it is precise without being pedantic, and it reads well most of the time. In the last two chapters it is occasionally less easy to follow the authors reasoning. This may be just because the multidimensional case is essentially more difficult to present. We should be grateful to Dr. Robinson for presenting us this useful book."]

Digital Foundations of Time Series Analysis by Enders A. Robinson & Manuel T. Silvia. Volume 1: Box Jenkins Approach (1979). Vol. 2: Wave-Equation Space-Time Processing.

Statistical Reasoning and Decision Making by Enders A. Robinson (1981) [Review by Paul C. Wuenschel in *Geophysics*: "The final section is a general discussion of the stock market, and the value of approaching investments with a view toward statistical chances of success."]

Time series analysis and applications by Enders A. Robinson (1981) [Review by Paul C. Wuenschel in *Geophysics*: "All chapters are worthy of close attention, demonstrating the author's flair for clear theoretical discussions and understandable examples. In short, this book has something to say on almost all aspects of time series analysis."]

Least Squares Regression Analysis in Terms of Linear Algebra by Enders A. Robinson (1981). [Review by William R. Green in *Geophysics*: "As another part of Robinson's recent publishing explosion, this book might tend to be overlooked in the sheer volume of his output (10 books in 4 years). This would be unfortunate, since it is an excellent treatment of the matrix methods that are so useful in modern data analysis. It is written to provide students of physics and engineering a grounding in applied linear algebra, and the author has succeeded admirably in attaining that goal. This is an excellent book worthy of consideration by any science or engineering faculty."]

Seismic Velocity Analysis and the Convolutional Model by Enders A. Robinson (1983] [Review by Professor Robert S. White, Department of Earth Sciences, University of Cambridge, Cambridge, England. "This is a splendid book intended for undergraduate majors in geology and geophysics. This book is for learning from and it teaches almost effortlessly. L ke any good teacher. Enders Robinson tells us the main points several times over to ensure that we understand them; yet he does so in such a way that one hardly knows that it is the same message. We can be grateful; that his teaching expertise is now accessible to us all through this book."]

Probability Theory and Applications by Enders A. Robinson (1985) [Review by Franz Thomas Bruss, professor of mathematics, director of the probability chair, at the Université Libre de Bruxelles, in *Zentrallblatt fur Mathematic und ihre Grezgebiete*, vol. 582. "The author stayed consistent with his aim to provide the student with a solid introduction to the understanding of randomness and, in particular, to give many examples and applications of a wide variety of situations. And for all sections that the reviewer has studied so far in more detail he finds, that this is nicely done. The book is an interesting alternative to other introductory texts on the subject, both in content and style."]

Seismic Inversion and Deconvolution: Part B. Dual Sensor Technology by Enders A. Robinson (1999) [Book review by Mrinal K. Sen in *EOS Transactions of the American Geophysical Union*, vol. 81, no. 32, August 8, 2000. "A book that specifically addresses the dual-sensor technology written by Enders A. Robinson, one of the pioneers of seismic data processing, is therefore a welcome addition to the geophysical literature. Although the book's high price tag will make it difficult for students to purchase a personal copy, every serious exploration geophysicists and research organization should possess one. This book is truly outstanding. I sincerely thank Enders Robinson for bringing us yet another classic."]

Time Series Analysis and Applications to Geophysical Systems (The IMA Volumes in Mathematics and its Applications by David Brillinger, Enders Anthony Robinson and Frederic Paik Schoenberg (2004}

Digital Imaging and Deconvolution: The ABCs of Seismic Exploration and Processing by Enders A. Robinson and Sven Treitel (2008)